The Game Music Handbook

The Game Music Handbook

A PRACTICAL GUIDE TO CRAFTING AN UNFORGETTABLE MUSICAL SOUNDSCAPE

Noah Kellman
With Foreword by Disasterpeace

UNIVERSITY PRESS

Oxford University Press is a department of the University of Oxford. It furthers
the University's objective of excellence in research, scholarship, and education
by publishing worldwide. Oxford is a registered trade mark of Oxford University
Press in the UK and certain other countries.

Published in the United States of America by Oxford University Press
198 Madison Avenue, New York, NY 10016, United States of America.

© Oxford University Press 2020

All rights reserved. No part of this publication may be reproduced, stored in
a retrieval system, or transmitted, in any form or by any means, without the
prior permission in writing of Oxford University Press, or as expressly permitted
by law, by license, or under terms agreed with the appropriate reproduction
rights organization. Inquiries concerning reproduction outside the scope of the
above should be sent to the Rights Department, Oxford University Press, at the
address above.

You must not circulate this work in any other form
and you must impose this same condition on any acquirer.

Library of Congress Cataloging-in-Publication Data
Names: Kellman, Noah, author.
Title: The game music handbook : a practical guide to crafting an unforgettable musical soundscape /
by Noah Kellman ; with foreword by Disasterpeace (aka Rich Vreeland).
Description: New York : Oxford University Press, 2020. | Includes bibliographical references and index.
Identifiers: LCCN 2020017864 (ebook) | LCCN 2020017865 (ebook) | ISBN 9780190938680 (hardback) |
ISBN 9780190938697 (paperback) | ISBN 9780190938710 (epub)
Subjects: LCSH: Video game music—Instruction and study. | Computer game music—Instruction and study.
Classification: LCC MT64.V53 K45 2020 (print) | LCC MT64.V53 (ebook) | DDC 781.5/4—dc23
LC record available at https://lccn.loc.gov/2020017864
LC ebook record available at https://lccn.loc.gov/2020017865

9 8 7 6 5 4 3 2 1

Paperback printed by Sheridan Books, Inc., United States of America
Hardback printed by Bridgeport National Bindery, Inc., United States of America

CONTENTS

Foreword • vii
Acknowledgments • xi
About the Companion Website • xiii
Introduction • xv

PART I GAME SCORING FUNDAMENTALS

1. Game Music Basics: Creating Emotional Arcs for Nonlinear Experiences • 3
2. Fundamental Game Scoring Techniques • 15
3. What Is "Game Music"? How Video Game Music Came to Be • 27

PART II CREATING THE MUSIC DESIGN

4. Player Immersion and Music • 45
5. Using Musical Codes to Enhance the Gaming Experience • 51
6. Exploring the Use of Musical Sound Effects in Games • 69
7. Music versus Sound Design: Defining the Sonic Relationship • 91
8. Designing Interaction between the Player and the Music • 103

PART III ADVANCED REACTIVE MUSIC CONCEPTS

9. Emergence: Music's Role in Player Experimentation • 121
10. Proceduralism: Using Game Data to Create a Real Time Score • 137
11. Advanced Algorithmic Music Systems • 149
12. Music in Virtual Reality • 161

PART IV IMPLEMENTING THE SCORE

13. Game Engines and Implementing Music in *Unity* • 177
14. Programming Crash Course for Composers • 193
15. Using Audio Middleware • 213
16. Essential Skills for Working with Developers: Version Control and Optimization • 231

Epilogue: Game Music Contracts and Navigating the Business • 247

Bibliography • 261
Index • 267

FOREWORD

Fruits of the Desert

Rich Vreeland

In visual media, there is a tendency to treat music like polish. Hired late in the process, composers typically begin working on a project that's otherwise nearly complete. Our task is to put the contributions of the rest of the team in a more flattering light.

In video games, we have a chance to cultivate a different way. With the prevalence of small, flexible, independent teams, and a growing stable of smaller, more creatively lenient publishers, we as composers can take on a more involved role. We can exist more like artists and collaborators, and less like the last-minute afterthought, taken for granted. We can stop being outsiders and start to embed more deeply with our colleagues. We can create more opportunities to earn their trust and respect. We can play a critical role in shaping the trajectory of the games we make.

There are a million and one ways to structure music in games. Take *The Secret of Monkey Island*, where the score modulates effortlessly from song to song as you travel the Caribbean. Or perhaps *Uurnog Uurnlimited*, where music algorithms evolve alongside player choices that other games rarely acknowledge. The scale of what's possible continues to grow year over year. And within this space are ways you and I can contribute that sidestep typical expectations.

Doing your best work generally requires time and freedom beyond the norm. You need more of both—so that you have the space to maneuver out of false starts and dead ends. The more you intend to challenge yourself and dance along the edge of what is possible, the more flexibility you need. You can't go out to the fringes without the trust of your coworkers, and you can't make your way back in a beneficial manner without sufficient time. So, if you can, get involved early, and don't forget to take breaks. Deep work requires deep rest.

If you do find yourself on board at the outset: Congratulations! Welcome to the desert. At this stage, everything is uncertain, and anything is possible. In this unfamiliar territory, you may not always have a "feel" for the trajectory of a project, and your colleagues may not either. There can be anguish in trying to write music for a game that has yet to figure itself out. You may find it fruitless developing a musical style in such foggy surroundings. It can, therefore, be beneficial to put off the content aspect of production until you have a better sense of the scope, scale, and essence of the project. For the moment, research and development may be the best things you can do.

I found myself in this position during the development of *Solar Ash Kingdom*. It took years to solidify the identity and mechanics at the core of the game. If I had spent much of that time making music demos, trying to hit a moving target, it would have been rather inefficient. It proved more fruitful to focus on weird, novel, and spontaneous ideas, workflow tools, and audio systems.

Solar Ash Kingdom was the first fully 3D game I ever worked on, and I treated the experience like being in an educational sandbox. Learning about tangential disciplines has a way of stoking the mind, and I learned a whole lot of math. I challenged myself to reinvent wheels, building things I'd seen other people do. I developed a tool for doing Doppler effects (how the pitch of a sound changes as you move quickly). I also built a dynamic reflection system, something I'd seen in AAA games, like Overwatch. But I tried to find new ways to use them. You could use dynamic reflections as a stylistic effect instead of a realistic one. What if Doppler pitch effects were used not just to alter the pitch of moving objects, but music? I became singularly focused on building out systems that could react in dynamic ways as you move through three dimensions. Being so engrossed in a new environment led me toward ideas for systems I'd never seen before as well. One example is a forest of trees that sing to you as you move through them. Your angle and velocity relative to each tree yield a different musical note. By diving so deep into sound propagation and vector math, I better understood the challenges and possibilities of this new paradigm, the 3D environment. And I came out the other side with ideas I would never have come up with otherwise, and a newfound ability to implement them. When you work in a more generalized way, you're more likely to create something with multiple uses down the road. I've already applied much of the new math I picked up while working on *Solar Ash Kingdom*. When I need a way to modulate a sound based on its positional relationship to other objects, for instance, I can take the dot product of two unit vectors. Whether you're personally familiar with these examples is unimportant. With time and space to play, you can collect an assortment of tricks, forever available to you and your peers (if you're generous), for the benefit of the game, and beyond.

There are ways to contribute besides music, though it can be challenging at times. You may be working with subject matter, people, or genres unfamiliar to you. I wouldn't have a great idea about how to contribute to the design of a fighting game, for instance, beyond giving feedback about whether it feels good or not to play. There are times when it's not a natural fit to chime in beyond musical boundaries, or perhaps you don't feel like it, and that's okay. Still, there may come a time when your unique background, skill set, and perspective afford you insights no one else has. This pre-production period is a great time to interface with new ideas outside of your comfort zone. When the pressure to deliver content isn't there yet, you have the opportunity to pick up new skills and learn new things from your colleagues.

We all have capacities that extend beyond our outward jxvi

specialty. I used to be a graphic designer. The sound designer I currently work with has also directed game projects. You might have an affinity for literature. You have more to offer than what it says on your business card or website. Getting involved early allows others to know you better, and gives you a chance to put more of your personality into the game. You get to be a part of the prototyping phase, whether you're exploring ideas in a musical silo or iterating on cross-disciplinary concepts with other members of your team. You, as much as anyone else, can theoretically steer the direction of what you're all making together. Sometimes even the smallest contributions, ideas, and suggestions can have an outsized impact.

Over time, there is a cumulative effect to all of this novelty exploration. You start to fill up your bag with hard-earned tricks, lessons from success and failure, and curiosities worth a look down the line. Some of these will be unique to games, while others may overlap with other musical forms, like film or theater. It could even go beyond into other mediums, like poetry or painting, should you ever find yourself there. For instance, in my bag, I know I can effectively create procedural sequences and iterations on many things (music, writing, etc.) using Markov chains. Or I can shift any note in a diminished chord down a half step to get to a dominant chord (and eventually to other keys). I know that a mono reverb in the middle of a signal chain can help to fill out a sound, which for me cross-pollinates with the way I use layers and effects in Photoshop. And by giving musical elements different bar lengths, one can create musical variations that go on without repeating for years.

The resourcefulness you accumulate in your creative travels may allow you to sidestep a mental block. I bring back old ideas from the dead all the time. The scores I've written for successful games like *FEZ* and *Hyper Light Drifter* contain tons of old, repurposed ideas. But I think we benefit from first stretching ourselves in those uncomfortable moments, continuing our search for new ideas. There are always new hurdles to cross, and if you want to do your best work, prior experience alone won't leap those bounds for you.

Early on in my career, I worried that sharing ideas and discoveries freely with others would dilute my uniqueness, jeopardizing my chance at success. But we all benefit tremendously from the collective knowledge accumulated by those who came before us. And even with an open spirit, sharing what you know with others, you will still have a bag of tricks wholly unique to you. There are some things we can't effectively externalize, and we all see things in a different shade. There's a little bit of the specific, a little bit of the broad, and a whole lot of you in every creative experience. Take some time to look through your bag, reflecting on what you've learned. Give your brain the time and space to make new cross-connections. It can help you make better sense of your work, your colleagues, and the broader creative world around you.

The groundwork for the "tried and true" approaches to video game music has been laid down and built upon for decades at this point. And you will find this book to be an excellent primer and guide to many of these techniques and concepts. But many exciting possibilities lie beyond the well-trodden path, in the endless desert of creative hypotheses. If you can, spend a few weeks in that desert trying out novel ideas: Explore an implementation approach that you've never tried before. Follow a silly thought experiment down a rabbit hole. Flip your usual strategies upside down. If you can stomach the struggle, the benefits you return with can be extraordinary.

Sure, you will sometimes find you've unintentionally reinvented the wheel. Or learn that you've failed at something someone else already discovered didn't work. But none of this exploration truly goes to waste. If you do uncover a gem, it may not suit the game. And trust me, when you come back from the desert, you may return with some inappropriate ideas! But you can always store them in your back pocket, for use at a later date. If your time was not overtly fruitful (to be fair, the desert doesn't have a lot of fruit trees!), you've at least learned something. And the more you wander, the more you'll learn the cost of your creative choices. You'll be humbled by what you could never accomplish without help. And you'll wise up to hairy problems that are hard to pull off even once.

Working on video games is not always fun and games (go figure). But creatively speaking, it has put me in some of the most compelling and confounding circumstances I've ever come across. So I guess what I'm saying is: don't be afraid to step into an oddly original landscape. You'll inevitably find yourself in a predicament anyway. So you might as well have a say about your surroundings.

Take this invaluable book, or a book like it, with you on your travels. You may enjoy breaking the rules more if you know what they are.

ACKNOWLEDGMENTS

First and foremost, I want to thank my friend Kent Kercher for the invaluable role he played in finishing this book as an editor and fact-checker, but most importantly as a great friend with a brilliant mind who offered countless opinions and nuggets of wisdom that helped shape the final product. I also want to thank Rachel Kercher, whose skill as an editor also helped us shape the text.

Second, I want to thank my wife Alba S. Torremocha, a wonderful composer who supported me throughout the entire process of conceiving this book, not only as a supportive partner, but also as a second set of eyes and ears which I trust unequivocally. Of course, my mother Jessie Kellman, my father Robert Kellman, and my brother Sam Kellman all played essential roles in supporting me throughout this process as well, and I will forever be grateful to them all. Sam's life experience with video games, combined with his skill as a writer, was particularly invaluable, and my parents were wonderful editors as well.

I would also like to thank Norman Hirshy for his guidance and encouragement throughout the entire process of writing this book, and for originally approaching me with the idea to do so.

It has always been of the utmost importance to me that this text contain the wisdom of many of today's greatest video game composers. Naturally, when I first contacted Rich Vreeland (*FEZ, Mini Metro*) and learned of his willingness to help me shape ideas and ensure that the book was as useful and beneficial to the reader as possible, I was beyond excited. I am so grateful to have had his opinions help guide me throughout the writing process, and I am also so thankful for the wonderful foreword he has provided for this text.

Even before beginning the writing process, I sat down to lunch with Jason Kantor, Audio Lead at Avalanche Studios, who offered me his advice and knowledge, as well as a variety of contacts who helped me build up my research. One of these people was Daniel Brown of Intelligent Music Systems, who took the time to walk me through his excellent procedural music system and then talk about many other game music–related subjects. He pointed me to composer Bobby Tahouri (*Rise of the Tomb Raider*) and lead audio designer Phil Lamperski (Crystal Dynamics), who each took time out of their busy schedules to discuss the music system of *Rise of the Tomb Raider* from their unique perspectives.

So many other wonderful people and skilled professionals offered their time and expertise, and I am so grateful to them as well. Composer Chris Velasco (*God of War III*) took the time to offer invaluable insight into his development as a composer both musically and from the business perspective, offering excellent ideas for my business chapter. Composer Lena Raine also gave wonderful opinions about designing reactive music, as well as enthralling conversation and insight about her acclaimed score to *Celeste*.

Some of my good friends, each highly skilled professionals, also offered me their knowledge. Composer and sound designer Mateo Nossa served as both a friend and as someone to run ideas by, someone with a top-level professional opinion that I value most highly. His skills in audio implementation proved invaluable for helping me shape the chapter about audio middleware. Mateo and I work together as part of the Frost Lab Studios team, and I want to thank the rest of the team, Stephen Judge and Julian Chelo, for their constant inspiration throughout the writing process as well.

Composer Mark Benis also contributed to the middleware chapter, discussing his stellar work on *The Worst Grimm Reaper,* and offering inside information about how he and the studio Moon Moon Moon used middleware to design the beautifully integrated music system. After noticing my use of *Mutant Year Zero: Road to Eden* (2018) as an example of music integration using middleware solution *Elias,* Mark introduced me to composer Robert Lundgren. Robert was gracious enough to walk me through the entire music system for the game, showing me the session and providing me with high-quality screenshots for use in the book.

I also want to thank Simon Ashby of Audio Kinetic, who offered invaluable advice about *Wwise* itself, and also pointed me in the direction of composer and music designer Guy Whitmore. Guy also took the time to meet with me, offering a great deal of valuable insight into the concept of Music Design and how *Wwise* (and middleware as a whole) can be used to create complex, yet smooth musical scores.

Sound designer Ronny Mraz and composer Zach Abramson are core members of the dream team that worked on Avalanche's *Just Cause 4* game score, and the two of them were an incredible asset in writing Chapter 9 on emergence. Their deep insight into the game's music system helped me to form a large portion of the chapter, and their work is a top-class example of both musical and technical ingenuity.

Sound designer Robert Rice was gracious enough to read my chapter on music in VR, using his vast experience with the medium to help me shape the chapter in a way that would be beneficial to the reader.

My close friend Harpreet Athwal, who has both a brilliant mind for programming and a wonderful knack for education, was of great assistance in forming this book's introduction to programming for composers. Harp always knows how to make a complex concept completely approachable and understandable for someone who has only just been introduced to it. I must also thank Jack Kelly and Frank DiCola of Game Revenant, the team that Alba and I worked with to create *Where Shadows Slumber.* The two of them not only helped proofread, but more importantly, they were very open to my including detailed information about *Where Shadows Slumber*'s music system and our design and integration process.

Last, but not least, I want to thank the many others in my life who contributed knowledge and support, and helped encourage me throughout the undertaking of writing this book. I am so grateful to all of you.

All in all, it's safe to say that this book is the culmination of the knowledge of many brilliant and unique minds in the world of video game music and audio. I am so grateful not only to those with whom I had the chance to speak directly, but also to all of those whose work is reflected in these pages as text I have referenced, or games that serve as examples.

ABOUT THE COMPANION WEBSITE

www.oup.com/us/thegamemusichandbook

As composers, we often learn best when we get to see, or more importantly, hear, critical learning concepts in action. For this reason, this book's 16 chapters include a variety of audio and video examples for you to pour over throughout the learning process. These can be found on the book's Companion Website. It will be apparent that a Companion Website example is available when you come across the following symbol in the text: "▶(1.1)." This symbol indicates that you will find an audio or video example illustrating the concept you are reading about, while the number next to the symbol corresponds to the chapter and example number on the website. For instance, if you are reading Chapter 2 and see "▶(2.8)" after a section about *stingers,* you should proceed to the Companion Website and choose example 2.8.

INTRODUCTION

The air bites at your numbing face as you sneak carefully through a snow-laden forest. You hold your bow at the ready in case danger appears in a split second. The tension of the moment leaves you hardly aware of the soundtrack that seems to be subconsciously playing in your mind: right now, a blend of light percussion that enhances the tension while still leaving you space to think as you approach an enemy camp. The light tribal drums seem to align with the rhythm of your footsteps as you sneak through the ominous brush, but their lightness doesn't penetrate the silence too harshly as you listen for the slightest sound of danger (Figure I.1).

Suddenly, through the dreary black, white, and dark green palette of the winter forest, you spy an armed man standing alone but alert—most likely, waiting for you to appear. The drums are aware. They react to his presence in your sights; a low, tribal rhythm grows louder and greater in intensity. You raise your bow. Your mental soundtrack follows you, the drums picking up slightly with denser rhythmic content that puts you further on edge. As you finally strike, a percussive cymbal hit sounds just as your arrow hits the man square in the face (Figure I.2).

The danger has temporarily subsided, and your mental soundtrack pauses. You enter briefly into a stealthy sneak, and light percussion picks up again. The music follows, again growing in intensity as you approach a group of enemies. Suddenly, the enemies spot you, and a low drum is struck loudly as you are forced to run, sliding behind a short stone wall to avoid being hit by gunfire. You are now in a state of full-out combat. Your heart pounds as low drums play a dense rhythm with high-velocity hits.

Somehow, you are able to wait for just the right moment to pop out from behind the wall, landing arrows in the two enemies you are battling. Just as an arrow strikes the head of an enemy, a high percussive timbre sounds, perfectly on cue as part of the drum beat itself. With a sigh of relief, you step out and lower your bow, knowing the danger has passed. The music subsides back to a relaxed state with more space in between notes, higher percussive elements, and ambient strings combined with synth timbres (Figure I.3).

The scene described in the preceding is a common combat sequence from the triple-A game *Rise of the Tomb Raider* (2013). The game, by Crystal Dynamics, features a musical score by Bobby Tahouri, all realized in real time with the use of software by Daniel Brown called the Dynamic Percussion System. This music system, implemented by lead audio designer Phil Lamperski, is a wonderful example of how detailed interactive music can be in games: it reacts on the spot to even the most minute player decisions, developing differently depending on each player's playstyle and how it informs the designed gameplay.

We will further discuss the implementation of the *Rise of the Tomb Raider* music system in greater detail in Chapter 10 as part of our discussion of procedural music systems; however, suffice to say that the preceding example illustrates just how different scoring a game can be from scoring any other medium. Why? Simply put, because

Figure I.1.
Lara Croft (controlled by the player) sneaking through the woods. *Rise of the Tomb Raider*, game disc, developed by Crystal Dynamics (Tokyo: Square Enix, 2015).

Figure I.2.
Lara raises her bow, preparing to strike an enemy as the music follows the on-screen action.

Figure I.3.
Lara battling two enemies as low drums match the intensity.

everyone plays games differently. Realistically, there is no way to predict how each player will navigate a game. This means that the final form and shape of our music can only be partially predetermined—the end result is, rather magically, the combination of our composition with the player's personality and playstyle. This is perhaps the most wonderful part of being a game composer—knowing that your score is not only bolstering the art of the game, but also having a direct relationship with the player.

With this in mind, as game composers, we must write differently, accounting for unpredictability and *nonlinearity,* and we must have a set of skills unique to the art of game composition in order to accomplish this.

Thus, we're going to go on an adventure through the colorful and infinitely fun world of game music. We'll learn everything, from how it all started, to the many fantastic ways in which game music is created today. The best part is, we'll learn it all by exploring many of your favorite games, as well as some games you may never have heard of. There are many different types of games out there, and each of them requires its own approach to creating the music and audio. The aim of this book is to keep things clear and concise, giving you the information you need to understand each concept while also providing you with fun and creative examples to open you up to the many compositional possibilities when it comes to writing game music. Our main goal will be understanding key concepts, both conceptually and technically, which are fundamental to being a good game music composer. Throughout the course of reading this book, you will gain not only key technical skills, but also an understanding that allows you to be creative and open-minded when designing the score for each video game project. Some of the skills you will learn are purely technical, like understanding programming and how to use middleware software, while others are more conceptual, like creating a music design that plans how the player will interact with your music, the relationship between the music and the other audio in the game, how the music will strengthen immersion, how the music will interact with the rest of the game world, and more—sometimes before ever writing a single note.

This book is divided into four parts. Part I: Game Scoring Fundamentals (Chapters 1–3) contains all of the most fundamental information you need to get started working on games. Chapter 1 provides a crucial introduction to scoring for nonlinear timelines. In it, we explore the art of creating emotional contour in our musical compositions even when the storyline or gameplay occurs out of order (i.e., nonlinearly). Chapter 2 dives right into the fundamental scoring techniques you need to understand to begin game scoring immediately. With Chapters 1–2 under your belt, you will be able to conceivably begin your game scoring journey already. Chapter 3 takes a deep dive into the history of game music and why its unique sonic aesthetics and form structures became standard—in other words, it gives you the info you need to "know your history."

Part II: Creating the Music Design (Chapters 4–8) focuses on creating a music design for your score. You will be provided a great deal of important information about how to make your score as immersive as possible, including how the relationship between the music and sound design (the game's other audio) affects scoring choices, using musical codes to elicit a specific response from the player, creating musical sound effects, and different ways in which the player can interact with the music.

Part III: Advanced Reactive Music Concepts (Chapters 9–12) dives into the forefront of game scoring, talking about advanced concepts like complex modular scoring, procedural music generation, algorithmic music in games, and writing music for virtual reality. Each of its chapters covers techniques that are growing more important and prevalent daily.

Part IV: Implementing the Score (Chapters 13–16) is the most technical of the four parts. Chapter 13 discusses game engines, with a particular focus on *Unity*, one of the most popular game engines of today. Chapter 14 is a programming primer written to be as accessible as possible for composers who want to pick up programming, a skill that will indubitably be invaluable to any game composer who takes the time to learn even just the basics. Chapter 15 discusses middleware, an important bridge between composing and programming that allows modern-day composers to create complex musical structures without needing to write complex programs. Chapter 16 covers two important subjects: version control and optimization. Each of these skills is too often neglected in game music education, but each is absolutely essential to any composer to wishes to work on an indie game.

Finally, the Epilogue offers a comprehensive guide to navigating the game music business, with a particular emphasis on being an indie composer. It has never been easier for a small band of people to come together and create a game, so naturally, the indie game industry is booming. This chapter will give you a number of places to get started in the indie game business.

I have been fortunate to work on some wonderful projects, and I have sometimes done so as part of PHÖZ, a game audio company that consists of fantastic composer Alba S. Torremocha and myself. One of the games we will later discuss is *Where Shadows Slumber*, a 2.5 mobile puzzle game for which we designed and created all of the audio.

When speaking about *Where Shadows Slumber,* I will sometimes refer to "we," and unless otherwise clarified, this will mean PHÖZ.

Before we dive into your game music education, you should know that this book was in many ways a team effort. It contains the knowledge and wisdom of many of today's greatest game composers for both indie and AAA games. Their being forthcoming and taking the time to meet with me for interviews was absolutely invaluable, and so, the knowledge you will be gaining is not coming just from me; rather, it is from many of the most creative minds in the world of video game music and game audio.

With all of that said, I think it's about time. Lets begin your journey into the magical world of video game music.

PART I
Game Scoring Fundamentals

1

Game Music Basics
Creating Emotional Arcs for Nonlinear Experiences

Introduction

Game music is often as unpredictable as the players themselves. Unlike any preceding musical genre, its core elements are dynamic,[1] reactionary, and interactive. One of the most attractive features of games is that players have the opportunity to transcend themselves, experiencing alternate realities firsthand, changing these alien worlds through action and choice, unlike the experiences offered by most media. In story-based games, players have a crucial influence over the narrative, creating individualized timelines by making personal, professional, political, spiritual, and other choices imperative to their formation. In games that are not story-based,[2] gameplay is still incredibly dynamic, the game engine reacting to player decisions at a detailed level that impacts the level of intensity, order of events, and other aspects of gameplay.

Scoring games is, needless to say, a detailed process that requires not only technical skills, but also the ability to empathize with the story, its characters, and how these will affect the player throughout gameplay. While scoring a film, one must be exceptionally privy not only to what is occurring on screen, but also to what the intended sentiment of each scene, movement, or facial expression actually is. The music must strengthen these goals without overpowering the narrative. In games, this is true as well; however, scoring emotion means creating a music system that can intelligently react to each player's actions, rather than simply creating a linear composition. In many cases, the music system is more or less collaborating with players, piecing together a symphony in real time.

Thus, game music is a wonderfully complex art form because each player may spend time in the game differently—exploring locations in random sequence, encountering enemies leading to incalculable combat durations, and, all in all, rendering the length of each play session variable. Our resulting compositions themselves become representations not only of our own vision or that of the game developers, but also of the players. Of course, accomplishing such a lofty goal cannot be achieved solely with traditional scoring methods. The fluctuating nature of gameplay and nonlinear storylines means that composers need new conceptual methodologies and technical tools. This chapter will introduce you to the fundamental approaches for composing game music. We will examine the concept of composing for nonlinear timelines and the challenges it presents, exploring how game composers can create emotional arcs for unpredictable storylines and gameplay without the foreknowledge of their length and order.

Creating a Music Design

Nonlinear composition is uniquely challenging, and incredibly fun because of it. As nonlinear composers, we don't know which places players will explore in what order, how long they will spend in each area, whom they will encounter, at what point they will engage in combat, how long that combat will last, what their play style will be (will they focus on stealth, taking out enemies without ever being noticed, or will they go in "guns blazing?")—the list goes on. We are forced to think ahead, composing for possibilities rather than absolutes. Since musical form is dependent on each gameplay, we cannot predict the final form of our composition, even when planning it out ahead in depth. Therefore, we have no choice but to become more than just composers. We must be *music designers* as well, creating not only the basic compositions, but also constructing systems that will determine how our music will function and interact with the player in the game, all before we ever write a note.

In his book *Composing Music for Games* (2016), composer Chance Thomas refers to a *music design* as "a comprehensive plan which outlines the purposes, tools and logistics for all uses of music within a video game." He points out that by having such a plan early in the process, "most features and problems can be thought through and resolved well in advance."[3] This is an apt description of the purpose of a music design. Coping with the unique challenges of game scoring requires a great deal of forethought. You cannot simply begin composing to what's on screen—you need to sit down with the development team to have a detailed conversation about what the music needs to achieve, how the player will interact with it, the relationship between the music and the sound design (sound effects and ambience),[4] the technical limitations of the game and its audio system, how much music is needed, and sometimes other elements. All of these questions are taken into consideration, and a *music design* is created.

Naturally, it follows that in order to create a strong music design, composers must have excellent prior knowledge about the technological side of game music composition. While some composers focus only on writing music, those who take the time to learn the fundamental *technical* skills of game scoring are able to play a key role in determining how the music of a game fits into the game design. This inevitably shapes not only the score, but even the game itself, in many unique ways. Composer Guy Whitmore, who spearheaded the innovative score to *Peggle 2* (2013), by PopCap Games, affirms this point of view: "What I resist is the idea of saying 'well, I don't do the technical stuff.' This [technological side of game music scoring] isn't technical, it's creative. That's why I like the term *music design*. It carries with it both the creative side, and the technical."[5] Indeed, when it comes to scoring games, the creative and the technical go hand in hand. Understanding both will give you the tools you need to be an invaluable asset to any game development team.

Diegesis

Diegesis is an imperative concept to understand when discussing video game music. A sound is *diegetic* when it exists within the game world (i.e., sounds like footsteps, characters speaking, and gun shots are all sounds the *player character* [the character being controlled by the player] would hear in his or her world). Sound is *nondiegetic* when it exists outside of this world; for example, a musical underscore is almost always nondiegetic, as it can be heard only by the player, but not by the character in the game.

To further clarify the distinction between *diegetic* and *nondiegetic* sound, imagine this scenario: a game character standing inside a living room walks over to a stereo system, turns it on, and begins to dance along to the beat. Within this scenario, the room is now filled with the music coming from the speakers of the stereo system; moreover, the character clearly hears and responds to that music, as he is dancing to the beat. In other words, the music coming from the speakers exists within the same world as the character; thus, this music is *diegetic*.

Now, imagine this scenario: two characters are on a romantic walk beside a still, deserted lake. As one leans in to kiss the other, the violins swell, reflecting their passion—yet there is no full orchestra hiding in the surrounding woods, nor have the characters brought along a boombox. In this scenario, the characters clearly do not hear this romantic music, as there is no source for it in the in-game world; the music is meant for only the player to hear. As this music exists only in the real world, and not the in-game world, this music is *nondiegetic*.

As we will see in following chapters, understanding the difference between diegetic and nondiegetic sound is particularly crucial for creating an effective music design. You will need to make decisions about which sounds in a game are diegetic, and whether you want these sounds to be musical or not. Remember to refer back to this section later on if you need a refresher on diegesis. For now, let us continue our discussion of nonlinear music composition.

Creating Emotional Arcs for Nonlinear Timelines

Even nonlinear games need to have an emotional arc, whether it be the arc to the entire game's story, or just a five-minute spell of gameplay. Because many games do not have a clear, linear narrative to follow, it is often essential that the music provide emotional context for the player as the game progresses. A well-placed musical cue can remind a player of why a certain place, character, or moment is important, creating an entirely new meaning in a specific instance or reinforcing a player's connection to that person, place, or time. A confident, triumphant character theme, for instance, can become ingrained in a player's mind with a strong association to the story's hero. Using this same theme in a downtrodden, sad musical context is a powerful way to signal to the player that an important development has occurred in that character's story. There are a few key techniques to consider when scoring for nonlinear timelines. These can be used to provide the player with emotional context throughout nonlinear gameplay. Let us begin by discussing the first of these techniques, locational scoring.

Nonlinear Composition, Technique #1: Locational Scoring

Locational scoring is the technique of writing musical cues that play depending on where the player is. These cues generally reflect not only the visual characteristics of each place in the game, but also the emotional or contextual meaning of the player character being present in that location at any given moment. In order to write strong locational music, we must first determine the function each location serves and what it means to both the player and the player character. Is this location a safe place, or it is dangerous? Does the player come here to save the game and store items with full confidence of its safety (like a safe room in the otherwise perilous *Resident Evil* series, or a home world in *Spyro the Dragon* [1998]), or is it full of enemies who could attack at any moment? Next, where is it in the geography of the game world? Is it in the player's home world, or is it on a distant alien planet? What does it look like? Does the architecture reflect a specific culture or group? Finally, what is this location's significance in the greater storyline? Does it first appear early in the game? Can the player find it and enter easily, or does entering require overcoming a substantial challenge? Once this level has been beaten, will the player ever return, or is this location entered only this one time? If the player will indeed return, does the location change the second time, and what is the meaning behind this change?

After answering the preceding questions, we can start painting a picture of what each locational cue should sound like. If it is a safe place free of dangers, then the theme will likely be slower in tempo and lower in intensity. It might use rounder, more inviting timbres, as opposed to harsh sounds with high frequency content. The classic *Pokémon Red* and *Blue* (1998) hometown theme for "Pallet Town," the town in which the player begins the game, features a diatonic, lullaby-like melody supported with harmony that moves easily around its home key of G major, all at a relaxed medium tempo.[6] ▶(1.1) This instills a sense of comfort in the player. The *Pokémon Red* and *Blue* "Wild Pokémon" battle music, on the other hand, is fast-paced—almost twice the tempo of the "Pallet Town" theme—using brash sounds to outline a chromatic progression between C minor and C♯ diminished—two keys with a distant, uneasy relationship.[7] ▶(1.2)

We can also use musical codes[8] to highlight specific aesthetic, architectural, cultural, or even political characteristics of a place. For example, the music for "Koopa Troopa Beach," a race track set along the sandy perimeter of a tropical island in *Mario Kart 64* (1996),[9] features an instrumental timbre that closely resembles a wooden xylophone, congas, and an Afro-Cuban rhythm in the percussion and bassline. The wooden timbres accentuate the level's outdoor, natural essence, while the Afro-Cuban rhythms in the bass and percussion use a pre-established musical association with a specific culture to establish atmosphere and inform the player about the surroundings. ▶(1.3)

Scoring location is an effective method for establishing an emotional connection between the player and a place. Each location in a game is meaningful for its own reasons, and creating the right association early on grants us the power to use thematic development, changing the meaning of our musical themes depending on how each individual story progresses, regardless of the order in which events occur.

For cases in which the player returns to the same place on multiple occasions, there exist interesting possibilities for creating an emotional arc for a story. Let us first consider an example in which the meaning of returning to a place has changed for the character. Our character begins the game as a common resident of a small town who has yet to step outside the town gates. In this case, the locational music (a town theme) reflects that this location is home, the one place where the character feels truly comfortable (much like the "Pallet Town" theme). The music portrays this locational intimacy by being tranquil, inviting, and grounded, with largely diatonic harmony that never strays far from home. As opposed to using bright synthesizers with pronounced high-end frequencies, the synthesizers used have been crafted with round timbres, rich in their low and middle frequencies. The theme, seen in Figure 1.1, is composed in the warm and inviting key of E♭ major. ▶(1.4)

After a series of events lead the character on a great adventure through faraway lands, leaving him renowned throughout the world, he returns home with a different social status and a heightened internal confidence. The character's return in this new state of being should be reflected musically by creating a second version of the town theme, one that portrays an element of triumph and confidence. The following example (shown in Figure 1.2) demonstrates a "Triumphant Return Theme" based on the original theme from Figure 1.1. In this example, we shift up to the key of F♯ major and feature brighter timbres playing the melody. We also increase the tempo from 90 bpm (beats per minute) to 110 bpm, using a march-like rhythm to enhance the comparison to fanfare, a style historically associated with triumph. The second chord of the progression is shifted to a ♭VII rather than a ii, while at the end of the progression, a D-major (♭VI) chord replaces the IV chord, a full departure from the F♯-major tonality, providing a noticeable shift toward a more prideful tone. This example illustrates that while it may be impossible to determine when the player will finally arrive back in the character's hometown, we may very well have the knowledge that when he does, his arrival will be triumphant and he will be seen as a hero. This allows you to examine the later statement of the theme, determining how best to musically foreshadow within the context of the iteration first heard in the game. ▶(1.5)

At his masterclass presented on behalf of the Game Audio Network Guild in April 2013, composer Richard Vreeland (aka Disasterpeace) used renowned indie game *FEZ* (2012) as an example to discuss how to create a strong narrative with locational music in nonlinear games.[10] In *FEZ*, there are no consequences for exploring the different parts of the game in random sequence. The player begins in the protagonist's home village, a 2D world filled with residents unaware of the existence of a third dimension. "Compass" is the cue that plays when the player reaches the central hub of the game. While the game

Figure 1.1.
"Original Town Theme."
Composed by Noah Kellman.

Figure 1.2. "Triumphant Return Theme." Composed by Noah Kellman.

begins linearly, this moment marks the last instance in which the player visits the same location in the same order in each play-through. Disasterpeace, therefore, chose to write stable and uplifting music. The piece is rooted in major 9 voicings, always falling onto their root notes for stability. The music compels one to feel safe and secure in this central hub, which can always be returned to after dangerous adventures.

There is also a musical cue that plays when you first depart from your home village. Interestingly, this track is heard only once in the entire game. The purpose of this singularity is to highlight the fact that only once can you leave your home for the first time. Disasterpeace decided that if the player ever returns to this area, no music will sound at all. He points out that leaving home for the first time is often a profound and evolutionary experience for a person: "I saw this as an opportunity to go beyond just writing locational music. I saw it as an opportunity to try to inform the story of what's going on. So, what happens is, you hear the song once, and if you ever come back to this area, you'll just hear ambience."[11]

Sometimes, a location undertakes a new meaning without any change to the character whatsoever. Rather, the player returns to the same location to find that the *place* itself has changed or evolved, that its atmosphere carries a new connotation, or that its residents perhaps have changed in some way. In such cases, the evolution of the location may affect the character, particularly if the transformation was unexpected. Toward the end of *FEZ*, there is an area where the player discovers an ancient civilization. Interestingly, the player has visited this area earlier in the game, but it appeared to be ancient ruins in a state of decay. Later in the game, it is revealed that this was an advanced civilization of great prosperity, and the theme is revitalized as the player visits the same location at its prime. The discovery of this ancient civilization is an important peak in the game's story. Therefore, Disasterpeace foreshadowed it on multiple occasions, extending the locational foreshadowing throughout the early stages of the game. The revitalized theme is called "Majesty," and it is heard when the ancient civilization is discovered. The music features more harmonic movement and lacks resolution, while also literally having a certain "majestic" quality. According to Vreeland, "This is a clear high point in the story. It's supposed to feel like a huge discovery, like you finally found the thing . . . if you play the game, there are all these little mentions of it. There are all these symbols and things you can uncover . . . it's foreshadowed by music in other parts of the game."[12]

A location's atmosphere can also change when that location is taken over by new inhabitants. These inhabitants might bring with them a different culture or political affiliation, changing the mood of the environment. In his talk at GameSoundCon 2017, game music scholar Gregory Rossetti focused on town themes and how they evolve throughout games, pointing out that they often develop based on political affiliation.[13] Sometimes, a political group within a game commandeers a town, in which case a new version of that town's theme is stated, musically reflecting the change of political power, as well as any tension created by this development. This new version of the theme tends to include musical codes that reflect the culture or home location of this new political group, as well as the group's philosophy. For example, if the group is militant, the music might include codes that suggest a military music style, like snare drums playing a march-based rhythm.

All in all, scoring location is a valuable method for reflecting and even creating narrative development in nonlinear settings. By using it, you will find connections between the game's characters, cultures, and political groups that allow you to foreshadow the evolution of the story as the adventure unfolds.

Nonlinear Composition, Technique #2: Character Scoring

One tendency in many nonlinear games is for the player to take on the role of the story's protagonist. This means that regardless of the order in which certain areas are explored, we can be sure that the character will always be present, reacting to events and giving the player a window into the character's emotions by reacting to different experiences. Players are often not playing the game as themselves, but rather as a character with preformed beliefs and an individual personality that may even contrast with that of the player. It is important to note, then, that there is sometimes a disconnect between the player's intentions and the character's reactions to the results of those actions. To borrow the words of game music scholar William Cheng, "Pressing a single controller button can produce actions that far exceed our expectations and direct operations. One of digital gaming's pleasures lies in how avatars outstrip our command, even if only in an illusory fashion."[14] As is suggested, the player character provides us with a somewhat more predictable personality to use as an emotional compass. Even if the player's actions are wholly impossible to foresee, the character as designed by the developers may have a limited set of reactions, tailored to various player decisions and their consequences. Therefore, in many cases our best course of action is to follow the character's emotional development by writing and developing character themes. The first statement of a character theme occurs when the character first appears on screen, providing a musical snapshot of a personality in its current contextual state and providing a foundation for musical foreshadowing. As characters build experience and evolve in various situations, their themes sympathetically develop, offering a complimentary musical evolution.

In *The Last of Us* (2013), the protagonist and player character Joel experiences the tremendous loss of his daughter Sarah in the opening sequence of the game.[15] In this heartbreaking scene, Sarah's theme is heard pouring out against Joel's sobbing and pleading. The core plot of the game takes place years after this tragedy. Joel has clearly dealt with the loss of his daughter and the dangers presented by a decaying world by shutting people out and repressing his ability to love.

However, he is forced to confront his emotions when he becomes the sole hope for the survival of a girl named Ellie, who is about the age his daughter was when she passed. Joel struggles to deal with his affection for her and attempts to push her away as a defense mechanism. Nonetheless, their relationship develops and they begin to grow close. Ellie manages to break down the years of walls Joel has put up. At this point in the story, Sarah's theme is cleverly restated, bolstering the emotional impact of Joel and Ellie's closeness, suggesting that Ellie is becoming like a second daughter to Joel. In this case, the theme transcends its belonging to Sarah with a devastating emotional impact, becoming instead a daughter theme.

Character themes also provide us with an excellent opportunity to create connections to characters who are not currently on screen. The *Final Fantasy* series is riddled with leitmotifs (themes attached to characters, places, or ideas) that develop throughout the game as they are encountered under changing circumstances. For example, Sephiroth, the main villain of *Final Fantasy VII,* has his own unique theme, which moves forebodingly between D-minor, C#-minor, and B♭-major/D triads.[16] This theme is utilized on various occasions throughout the game in which Sephiroth has had an influence, stirring the player's anger with Sephiroth, indicating danger, and most importantly, connecting Sephiroth's influence to malicious circumstances in players' minds, even if the villain is not literally present at those times. At one point in the game, Sephiroth has begun a series of events that lead to characters Tifa Lockhart and Barret Wallace being tied up in order to be executed by gas chamber. This event is an indirect result of Sephiroth's summoning of the Meteor spell, or "The Ultimate Destructive Magic." As the execution is being carried out, Sephiroth's theme plays ominously and triumphantly, connecting his actions to their detrimental results, despite a lack of his physical presence in the scene.

Super Mario 64 is also an excellent example of a game that successfully uses character themes to build emotional connection to characters who are rarely seen. Saving Princess Peach is the core objective of the game's story, despite Princess Peach rarely being encountered in the game. Her theme is established in the opening cinematic, and is also a central focus of the pleasant music playing in the background whenever the player is in the castle, which serves as a safe central hub from which the player enters levels. When this theme is finally heard at the end of the game after Peach has been saved, the player's connection to her has been noticeably strengthened by the music. In his video series *Game Music Archeology,* game developer James Covenant discusses various examples of leitmotif in *Super Mario 64* and subsequent series. According to his analysis, "Princess Peach has even less screen time than Bowser. I would argue that her character is developed primarily through the music. In a way, all of the time the player spends in the castle listening to Peach's theme song is an opportunity to form an emotional connection to the character."[17] Indeed, when Peach is finally rescued, the hours of peaceful castle exploration have a strong impact, giving players a deep sense that they have fought and succeeded in rescuing someone who represents peace and tranquility in the Mario universe.

Nonlinear Composition, Technique #3: Gameplay Scoring

In many cases, games simply need underscore that reacts to the gameplay regardless of any narrative. Non-narrative scoring, or rather scoring gameplay, is an essential piece of the game music puzzle. Many games feature specific gameplay types that arise numerous times throughout the course of the game, and you will need to create musical tracks tailored to specific instances. Some obvious examples of gameplay scoring (also sometimes referred to as situational or event-based scoring) are musical cues designed specifically for combat, stealth, exploration, boss battles, and mini games, though many other types of events might require their own musical cues as well. The aforementioned *Pokémon Red* and *Blue* games, for example, use iconic battle music composed by Junichi Masuda, which tends to begin with a startling descending chromatic sequence each time the player is cornered by an enemy. Masuda wrote different battle themes depending on what type of battle it is, for example, "Wild Pokémon,"[18] ▶(1.6) "Trainer Battle,"[19] ▶(1.7) "Gym Leader Battle,"[20] ▶(1.8) and "Final Battle"[21] ▶(1.9) themes. Many open-world RPGs[22] require a great deal of exploration music to keep players company while they are traversing the landscape. Notably, *The Witcher III: Wild Hunt* (2015) is famous for its soundtrack by Marcin Przybyłowicz, which underscores the player's excursions through the seemingly endless game world.

Even if certain events exist only as inconsequential vignettes in a complex greater storyline, they still require their own emotional arcs in order to fulfill their immersive duties, and music is an essential part of creating it. In most cases, composers are unable to truly predict the arc of such gameplay, and thus reactive forms are created such that the music can evolve with the intensity. Many story-based games tend to have mechanics built in that require non-narrative scoring. RPGs like *Final Fantasy, Pokémon,* and *Golden Sun* have profound narratives. However, they also include combat music, which generally loops during battles. They also have unique combat music depending on the importance of a battle or who it is with. This is why in many cases, the score must either ignore the details, using locational scoring to create atmosphere and feeling for the gameplay; or, it must be highly detailed, accounting for multiple possibilities so that the complex, reactive system can accurately conform to gameplay. All in all, the scoring of gameplay is a unique musical form, and the result is a dynamic composition tailored perfectly to adaptive gameplay.

Competitive games serve as excellent examples of gameplay scoring and how it leads to unique formal development because they are often closer to sports than stories. From the composer's point of view, a multiplayer match is nothing like your typical narrative compositional challenge. The same type of match may be played over and over, making it difficult to write music that doesn't get repetitive. Like in sports, each match can vary in speed and intensity, largely because of different players' strategies and playing techniques. For first-person shooters series like *Halo* and *Call of Duty*, the music tends to be designed such that it can react to these varying levels of intensity, making it feel unique to each match. Composers also generally avoid overly recognizable themes within competitive matches, as well as frequency areas that interfere with the sound effects of the game. By following these guidelines, the music becomes imperceptible, complementary underscore, enhancing the competitive experience while staying out of the way of gameplay.

For brawl games (like the *Super Smash Brothers* series) or fighting games (like the *Soulcalibur* series), the music is quite the opposite, often being thematic, orchestral background music that functions as even more added stimulation to the player. In these cases, subtle underscore would be boring, whereas epic thematic music works quite well. This is likely in part due to the fact that unlike first-person shooters, these games do not strive for realism, thus making noticeable music acceptable and even fun. In any case, many brawl games feature expressive and magnificent orchestral scoring with soaring themes, almost like a game music version of a pop song. These themes recurrently get stuck in player's minds, creating strong associations with the fun experience of playing games themselves and leading them to enjoy the music outside of the games as well.

Conclusion

Whether you are scoring location, character or simply designing the musical cues and rule set for event-based scoring, it is still always important to remember that your entire score should follow a unified aesthetic, serving as a musical glue that helps hold the entire game together. Scoring location provides us with a strong platform for using music to create ambiance and atmosphere when the story or even the arc of gameplay is uncertain. Since places change based on in-game events, their associated locational themes can be developed. Having knowledge of a location's fluctuating circumstances or appearance provides us with the opportunity to create foreshadowing by using different versions of the themes. All in all, location scoring done well is an excellent way of gluing together different narrative developments throughout a nonlinear timeline.

While the player may be wholly unpredictable, game characters are designed by the developers and therefore have calculable reactions. Using our knowledge of the characters' personalities, we can score their reactions in relation to game events, essentially bypassing scoring the player's decisions in order to score how the character feels instead. Gameplay scoring is also a fundamental part of the scoring process, requiring the composer to create an emotional arc that emerges as the situation requires. All in all, nonlinear scoring is an art that requires preparation and thoughtfulness. It is also, luckily, incredibly fun. In the following chapters, we will continue to examine the unique challenges of game scoring, delving into technical and conceptual methodologies that will empower you to be an excellent game music composer.

Exercises

Rugged Road is a hypothetical nonlinear RPG that takes place today in New York City. The game is centered around an impulsive character named Robert. He is ruggedly handsome, headstrong, with a slight New York City accent, and he doesn't take no for an answer. However, seeing as he grew up in a rough neighborhood, he has a tendency to get mixed up in crime. He has a strong moral code, but he is willing to break it to protect those he loves. In the first scene of the game, the player is introduced to Robert and how to control his movements using the controller. Robert's home borough is Brooklyn, and his apartment serves as a safe house for the first act of the game. Whenever something has gone awry, the player can always go back there to store items and clothing, and take a breath of fresh air in the comfort of its safety.

At the end of Act I, Robert, controlled by the player, will have an important decision to make. His girlfriend Sheila is controlling and manipulative, but her father is also high up in the mob. If Robert stays with her, Sheila's father will soon pressure him to work for the mob, a path Robert will be unable to avoid without insulting her charismatic and manipulative father. Robert will be forced to dive deeper into a life of crime, putting him in morally ambiguous situations. Breaking up with her will have certain advantages, for example, bringing Robert closer to his dream of vacating New York City altogether. However, it will also trigger the mob to take over Robert's neighborhood, with henchmen roaming the streets trying to track him down. Suddenly, his apartment will become unsafe as it is constantly being watched. They will also take all of his possessions.

1. Create a character theme and multiple variations for Robert.
    a. Compose a character theme for Robert. Use this as an opportunity to reflect his personality—his strong moral code, confidence, and his inability to take no for an answer.
    b. Remember, in Timeline A, Robert will stay with Sheila, becoming engrossed in crime. He will feel morally torn apart inside, conducting business he feels is wrong in order to please Sheila and her father and to raise enough funds to get out of the city once and for all. Following this timeline will cause him a great deal of *internal* pressure and stress. He will lose confidence in himself and grow sad as the story goes on. Compose an iteration of Robert's theme that conveys his internal struggle.
    c. In Timeline B, Robert will break up with Sheila and he will feel strong morally, following his code. However, having lost his apartment, possessions, and being pursued by numerous mobsters, he will be under a great deal of *external* pressure. In this scenario, he does not face any internal demons concerning doing the wrong thing. Therefore, he feels happier and more confident despite the dangerous situation he is in. Compose a version of Robert's theme that portrays the external tension he feels due to the precarious situation surrounding him.
2. Create a "home" locational theme for Robert's apartment.
    a. The first version of this theme will play in Act I for all players. Compose this theme to give the player a sense of safety when in Robert's apartment. See if you can somehow foreshadow the later version of the theme that will play if Robert's apartment becomes unsafe.
    b. If Robert chooses to break up with Sheila, his apartment will become one of the least safe places he could possibly go. However, he will need to break in later to retrieve important possessions. As he walks into his trashed apartment, a variation on the "home" locational theme will sound. Compose this variation to reflect the emotion of returning home to find it unwelcoming and unsafe.

Notes

1. Karen Collins, *Game Sound: An Introduction to the History and Practice of Video Game Music and Sound Design* (Cambridge, MA: MIT Press, 2008), 4. Collins states that *adaptive* audio is that which reacts to changes in game state and game parameters; *interactive* audio is that which is directly interactive with the player (for example, a sound that occurs at a button-press); and *dynamic* audio encompasses both.
2. In referring to games that are not story-based, this means that they have little to no narrative, and/or that the narrative is inconsequential to the game's development and gameplay. There are many types of games that are reactive or interactive, but are not story-based, including action, puzzle, real time strategy (RTS), simulations, shooters, and educational.

3. Chance Thomas, *Composing Music for Games: The Art, Technology and Business of Video Game Scoring* (Boca Raton, FL: CRC Press, 2017), 57.
4. Music and sound design relationship types will be explored in great deal in later chapters.
5. Guy Whitmore, interview with author, April 23, 2019.
6. "Pallet Town" Theme—*Pokemon Blue*, game cartridge, developed by Game Freak (Kyoto: Nintendo, 1996).
7. "Wild Pokemon" Theme—*Pokemon Blue*, game cartridge, developed by Game Freak (Kyoto: Nintendo, 1996).
8. Musical codes are predetermined associations that the player has with specific musical elements, such as melodies, instrumental timbres, harmonies, or certain musical structures. They are discussed in greater detail in later chapters.
9. *Mario Kart 64*, game cartridge, developed by Nintendo EAD (Kyoto: Nintendo, 1996).
10. Rich Vreeland, "Philosophy of Music Design in Games—Fez," YouTube video, 48:31. April 20, 2013, https://youtu.be/Pl86ND_c5Og.
11. Rich Vreeland, "Music Workshop—FEZ," YouTube video, 44:22, November 20, 2014, https://www.youtube.com/watch?v=PH04VJ8jxvo.
12. Vreeland, "Philosophy of Music Design in Games—Fez," 2013.
13. Gregory Rosetti, "RPG Town Themes: Evoking Place and Cultural Identity through Music," GameSoundCon (convention lecture), Millennium Biltmore Hotel, Los Angeles, CA, October 9, 2018.
14. William Cheng, *Sound Play: Video Games and The Musical Imagination* (Oxford: Oxford University Press, 2014), 9.
15. *The Last of Us*, game disc, developed by Naughty Dog (Tokyo: Sony Computer Entertainment, 2013).
16. Nobuo Uematsu, "One-Winged Angel," on *FINAL FANTASY VII: Original Sound Track*, SQUARE ENIX, 1997, MP3, accessed December 13, 2019, https://open.spotify.com/track/02hiFgacAzz7zGl5EF53eo?si=u_K90TO3SKOx4hJvzkR_Dw.
17. James Covenant, "The Music of Super Mario 64 | Game Music Archaeology Ep. 1," YouTube video, 16:21, March 2, 2017, http://www.youtube.com/watch?v=PE7r5trvOUI.
18. Junichi Masuda, "Battle! (Wild Pokémon)," 1996, on *Pokémon Red & Pokémon Blue: Super Music Collection*, The Pokémon Company/OVERLAP OVCP-0006, 2016, compact disc.
19. Junichi Masuda, "Battle! (Trainer Battle)," 1996, on *Pokémon Red & Pokémon Blue*.
20. Junichi Masuda, "Battle! (Gym Leader Battle)," 1996, on *Pokémon Red & Pokémon Blue*.
21. Junichi Masuda, "Final Battle! (Rival)," 1996, on *Pokémon Red & Pokémon Blue*.
22. RPG is short for *role-playing game*. In RPGs, the player assumes the role of the character and greatly determines how that character develops by deciding many important factors. For example, the player determines the character's strengths and skills, political associations, and often even its personality as the player makes important decisions throughout the game's story in regards to treating certain situations.

2

Fundamental Game Scoring Techniques

Introduction

Locational scoring, character themes, and foreshadowing are strong methods for overcoming the challenges of creating emotional arcs when composing for multiple temporal possibilities. However, they provide only the first half of the music design, affording us a long-term picture of narrative development in games. Gameplay itself tends to be quite repetitive. Imagine the multitude of games that require players to fight an unforgiving boss, only to die many times and repeat the process over and over until they have finally mastered the skills necessary to achieve success. Music is an essential tool for relieving any potential frustration or monotony that results from the repetitive nature of gameplay. The tools available to composers today provide us with methods for creating dynamic scores that react to momentary changes. When used well, musical development becomes a subtle reaction to the events on screen. Indeed, the variable nature of gameplay is also evident in the uncertain lengths of each play session. A particularly skillful player might defeat a boss in minutes, while another might repeat the process for hours before succeeding.

Consequently, we cannot simply write a piece of music from start to finish and expect it to time out correctly with the gameplay. Composers generally pre-render[1] (or bounce) our music, so we have to think ahead about how we can make it adaptable to variable game lengths. When all is said and done, we must use techniques that account for nonlinearity, reacting to quick changes in gameplay, while also repeating music since it is generally nonsensical or downright impossible to compose such a multitude of music that repetition is completely unnecessary for an experience that can last for an uncertain amount of hours. However, music also has the potential to increase monotony if it becomes too repetitive itself. In this chapter, we will delve into the technical aspects of composing nonlinearly, discussing the common techniques used in games today and empowering you to use these for your own game music composition.

Game Scoring Techniques

It's a clear, bright day. Just the sort you dreamed of only months before as you lay curled up alone, seeking solace during a cold, dreary winter that seemed it would never end.

How fortunate you feel to have put the winter behind you as you peer down the promising, sunlit path ahead. You hum as you sit comfortably atop your noble steed and best friend. You can hear a soundtrack to your life playing in the background: the oboe sounds an uplifting melody, while the strings provide a motor that helps propel you forward, almost in sync with the footsteps of your horse. You nap, dozing in and out as you travel, not noticing the sky slowly growing darker. Through half-open eyes, you notice a strange red hue over the clouds and a dark, gothic structure in the distance. Your mental soundtrack becomes more dissonant—a tense, violin tremolo subtly fades in to occupy the high register, almost like a painful buzzing in your brain. It seems to grow ever so slightly louder the closer you get to what now appears to be a large, abandoned cathedral. You arrive at the building to find that the door is halfway off its hinges, revealing a pitch black darkness within. The sky is now blood red, and you know you have no choice but to enter. You feel both terrified and yet somehow also seduced as low strings play atonal glissandi and a female choir sings a dissonant melody, like a group of beautiful sirens calling you inside. . . . ▶(2.1) End scene.

Imagine the preceding is an open-world RPG (role-playing game) that you have been tasked to score. In this medieval adventure which we will call *Warlock Beguiled*, you play as a handsome and heroic, 200-year-old knight with magical abilities. You are tasked with finding out who murdered the king of the realm. The current quest involves locating and entering an ancient, enchanted cathedral to find the witch who created the very poison used to kill the king. First of all, you don't know when the player will begin this quest. The quest itself doesn't actually activate until the player enters the cathedral. Players could, conceivably, ride around in the area surrounding it without even fully realizing they are nearby. Soon, we will discuss how the music in the preceding description was functioning and create an abridged *music design* together for this nonlinear game using a technique called *vertical layering*. In order to create vertical layering, however, we must first understand the concept of looping.

Looping

The single most basic and effective technique for achieving seamless musical repetition is looping. *Looping* involves creating musical cues that repeat smoothly without repetition being noticeable to the player. As the gameplay calls for changes in the music, loops *transition* between each other seamlessly. If well executed, a loop allows the music to exist underneath the gameplay, "looping" at the end of each track, all without any perceptible indication of musical repetition. Accomplishing this level of smoothness requires two core elements: First, in order for a loop to be unnoticeable, the composition itself has to be musically clever, and second, the bounce has to be created properly at the technical level. Try following these general guidelines to make your loops smoother and less perceptible:

- Avoid blatant musical elements that make the looping of the composition obvious. In his book *Composing Music for Games,* Chance Thomas points out that using a recognizable theme in a loop can have the unwanted effect of rendering the loop quite obvious to the player. This is because each time the loop comes around, the player will notice this theme, making the repetition blatant and the gameplay less immersive.[2] An obvious musical device placed exactly at the looping point, or end of the loop, can also have a similar effect, drawing the player's attention toward the music and thus away from the game. For example, a loud, transient drum hitting at the end of a beat or directly at the beginning or end of a loop would make the looping point quite obvious. As a rule, place looping points where they will be the least conspicuous. In our companion site example, a track from the 2019 mobile game *Grobo,* see if you can hear where the form loops. In "Version 1," this looping is fairly inconspicuous, using a quick *crossfade* placed directly on beat one. In "Version 2," an unnecessary drum fill at 01:08 highlights the end of the loop, drawing attention to the fact that the track restarts. Since the player will likely only be partially paying attention to the music, whether or not the fill is musical is somewhat beside the point. Rather, what is important is that this fill is unnecessary and conspicuous. ▶(2.2–2.3)

- Use multiple formal sections with differing styles, instrumentation, or contrasting emotional content. Following this method carries players through the composition seamlessly. They lose touch with each previous section naturally as a new musical world envelops them. By the time the music returns from the second section to the first, the first is no longer the primary inhabitant of players' short-term memory.
- When possible, try looping during an ambient section. As we have discussed, the loop should occur at a point in the composition when it will be least noticeable. If there is an ambient or arrhythmic section of the cue, this often serves as an excellent opportunity for creating an imperceptible loop. The music simply flows through the ambience as if no change occurred at all. In our companion website example, the loop occurs somewhere between 01:50 and 2:03. Can you hear it? ▶(2.4)
- Avoid any clicks in the loop. Particularly for rhythmic compositions, it is important that the loop occur exactly on a beat with a sufficient *crossfade*[3] to avoid any clicking. On the companion website, you can hear an example taken from PHÖZ's score to *Where Shadows Slumber* (2018), purposefully looped without following these guidelines. At the end of the loop, there is an audible audio click, which occurs because no crossfade was applied. The looping point is also quite obvious from the musical standpoint because it occurs somewhat randomly, out of time, and without the use of any smoothing techniques. ▶(2.5)
- Make sure the beginning of your loop includes any audio tail from the end of your composition. If you have any reverb or delay in your music whatsoever, the loop should be bounced such that the tail of the end of the piece is overlaid across the beginning of the track, therefore making the loop feel natural. One method of accomplishing this is to simply paste your loop three times, create crossfades on each side of the middle region and then bounce your loop from the center of each crossfade. In today's world, middleware software[4] makes looping and transitioning between music segments quite seamless. It also makes bouncing easier, since rather than bouncing perfect loops, you can simply bounce your music like a regular track and set up any looping, crossfades, and transitions in the software. We will dive deeper into middleware in later chapters. In our companion website example, note the delay throw at 01:25, which seamlessly continues at the looping point, making the loop imperceptible. ▶(2.6)

Transitions

Loops generally must be capable of switching from one to another at any given moment. We use *transitions* for this reason. The creation of transitions is a technical skill in and of itself. As game composers, we need to think ahead about how each track will transition to each other possible track. Technically, a simple crossfade between two tracks is a basic form of transition, but often they are more complex, requiring a musical *transition segment* that plays in between two cues to get from one to the next.

Transitions Types/Timings

- *Immediate*: In the case of an immediate transition, the music will instantly change from one track to the next. Generally, an immediate transition requires a short crossfade to avoid any clicks.
- *On the beat*: This transition type will happen directly on the next beat of the music. Middleware software makes this transition type fairly easy to achieve since it maintains the tempo of the music on an internal clock within the software. This transition type is excellent for situations when you want to keep the transition in tempo, but you also need it to happen quickly—in other words, a situation in which there is not sufficient time for the music to wait until the end of the bar to transition.
- *Bar*: Sometimes, you may have ample time and prefer to keep the music transition as smooth as possible. In such a case, you can make the transition happen at the end of the current or next bar of music.
- *Phrase*: To be even smoother compositionally, you may prefer to have the transition occur only after a musical phrase has been completed. The downside to this method is that if you have long phrases, it might take longer for the transition to take place, making it more conspicuous if the on-screen action transitions more quickly than the music. However, this works great for situations when you want the music to be utterly seamless and exact timing is not a primary concern.
- *Custom markers*: This simply means that the music transition will take place at whatever custom location in the track you have selected. For any case in which the previous transition types won't work, a custom marker may solve the issue.

Transition Segments

As discussed before, a *transition segment* provides us with a method for creating seamless musical transitions when a simple transition type will not suffice. There is a variety of cases in which this is particularly useful:

- *Tempo changes*: When dealing with a tempo change, any quick transition, whether immediate or even at the end of a phrase, can become quite noticeable. To solve this issue, one should consider creating a transition segment containing a short musical block or phrase that includes an accelerando or ritardando to carry the composition from one tempo to the next. In our companion website example, the transition segment begins at 00:24 and carries us into a new tempo, staring at 00:40. ▶(2.7)
- *Time signature changes*: While changing time signatures directly may sometimes be smoother than changing tempos, there are certainly cases in which a transition segment will serve as an excellent tool for making the change feel effortless.

- *Key changes*: Moving from one key to the next generally requires transitional material. In such cases, transition segments make this process quite musical.
- *Other*: Transition segments may also be useful in cases where you simply wish to include transitional material for the sake of the composition and its development itself, even if smoothness isn't an issue.

Stingers

A *stinger* is a short musical sound effect that plays on top of the music. ▶(2.8) Stingers have a variety of common functions.

- *Smoothing transitions*: Stingers are often used as an extra element to smooth transitions, particularly if the music system being used does not allow for transitions in tempo. By playing a stinger on top of the music, any lack of smoothness in the transition can often be covered up. Note how our stinger helps with the transition in our companion website example at 00:26. ▶(2.9)
- *Musical devices*: Stingers are also quite useful as a musical device. For example, in horror games, which employ sudden musical changes when one is being attacked or danger is nearby, they are used quite often as a surprising musical element that enhances the jump scare for the player. ▶(2.10)
- *Musical signals*: In *Understanding Video Game Music,* Tim Summers points out that "[s]tingers can act as musical signals to the player and frequently challenge the distinction between sound effect and music by retaining a musical quality, but being deployed with an immediacy and anchor in the game world that is similar to a sound effect."[5]

One technical consideration to keep in mind is that since a stinger is generally overlapping the music, there is high potential for the volume to exceed its maximum threshold and "clip" or distort.[6] You should be aware of this during both the composition and mixing processes to ensure you avoid this mistake.

Using Control Input to Affect Music

Any game you play receives input from outside and reacts to it. When you press a button, a signal travels from your controller into the game and causes something to happen. Remember, music is stored in files. The game triggers a file to load based on something in the programming, and often that programming is responding to the player.

Control input, the data being sent to control game or music parameters, is an important concept for composers to understand because we can use it to control various aspects of our music system. Just as the game receives control input from a handheld controller, a music system receives control input from the game itself. In his book *Writing Interactive Music for Video Games,* Michael Sweet describes a variety of control input elements that can affect the music system, including character location, enemy location or proximity, environmental effects like weather or time of day, the level of suspense in a scene, any emotion that the player should be feeling, the health of the player, non-player characters (NPCs) that may interact with the player, the action that the player is currently attempting, and how many puzzles the player has solved.[7]

Understanding that you have the power to design your music based on any of these control input elements, and more, should open your eyes to the world of possibility in which interactive music thrives. As you will soon understand, *horizontal resequencing* and *vertical layering* are both game music techniques that interpret control input data and react to it musically. However, each method reacts differently. In the end, it will be up to you to understand all of the techniques described in this book and to then creatively apply them yourself, depending on the requirements of each game you work on.

Game State

Before diving into creating our music design, let us go over the term *state*. In game design, all of the current parameters, mathematical values, positions of objects, and other values attached to the objects in a game at *this* very moment in time are essentially what make up your *game state*. For example, if your character is swinging his sword, he is in a different state than if he were simply holding it by his side. If he has low health, he is in a different state than if his health were full. Programmers, however, more frequently use the term to signify a specific circumstance the game can be in at any time, or in other words, a grouping of settings that defines the larger style of gameplay at a given time. For example, they might refer to a "dead" state that the game enters when the player runs out of health and dies. They might also refer to a "respawn" state, the state the game is in during the brief interval of time it takes the game to revive the player, or even something like "rampage," when the player has picked up an item that gives him super-strength abilities.

A *finite state machine* can be thought of as what controls the transitions between a predetermined group of states.[8] For example, a finite state machine might be composed of the states "exploration," "stealth," and "battle." The game would be in an "exploration" state whenever the player is simply wandering around. "Stealth" would be reserved for when she is near an enemy but the enemy has not yet detected her. Finally, "battle" would be for when the player is actually fighting an enemy. The state machine would allow the programmer to create specific transitions, depending on which state "A" is transitioning to the next, "B." "Battle" would have a different transition to "stealth" than it would to "exploration," for example.

As composers, we use the state machines created by the developers to transition between different musical cues. From our example, we can immediately see that *Warlock Beguiled* has an *exploration state*. While our warlock is simply riding through the forest, not necessarily engaged in any particular activity other than exploring the landscape, happening upon people and objects, and so on, the game can be said to be in an exploration state. This means that it will require exploration music. Some players might simply explore for hours, so we must be careful to design our music such that it doesn't become overly repetitive and boring. Rather, the music should encourage exploration, making it feel more fun and less monotonous. Let's decide that our exploration music will a) reflect the game's landscape using instrumental timbres, b) convey the relaxed, adventurous feeling of exploring by using mid- tempo rhythmic elements, and c) loop seamlessly so that repetition is not noticeable.

Horizontal Resequencing

Horizontal resequencing is a method of cueing up linear slices of music in an order that is determined by player input. In other words, pieces of music play from left to right without any noticeable transitions. It largely revolves around looping, transition segments, and stingers. In this basic diagram (Figure 2.1), the music moves horizontally from left to right, transitioning to the next cue when the player engages in battle.

In reality, however, horizontal systems tend to be more complex, accounting for multiple possible timelines. These possibilities are, of course, predetermined by the development team and the composer. In Figure 2.2, there are three possible directions in which the music could head. Depending on what the player chooses, the music will react and transition horizontally to the appropriate cue.

Horizontal resequencing is, overall, an excellent method for dealing with nonlinear gameplay. However, it is somewhat limited in what it can achieve. For example, what if instead of changing to a completely different cue, we want to develop the current cue, adding or subtracting different instruments depending on the intensity of the gameplay, or the different actions the player is carrying out? The next technique we will discuss is referred to as *vertical layering,* and it allows us to accomplish just that.

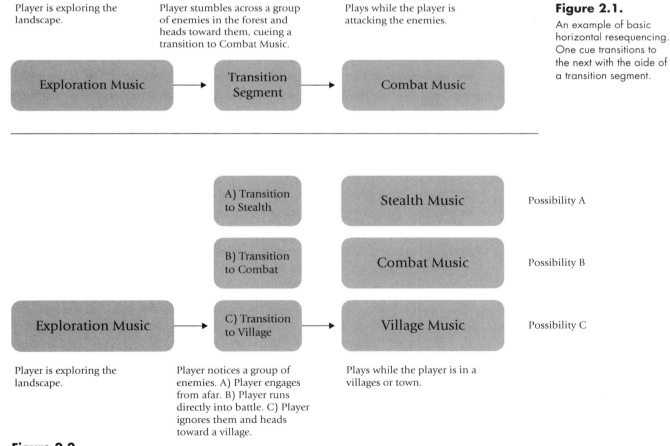

Figure 2.1.
An example of basic horizontal resequencing. One cue transitions to the next with the aide of a transition segment.

Figure 2.2.
A more complex horizontal resequencing.

Vertical Layering

Let us finally return to the compelling gameplay of *Warlock Beguiled* to design a music system using vertical layering. You might have taken note of the fact that certain instrumental *layers* grew in volume, depending on the player's proximity to certain game objects. In this case, the dissonant violin texture grew louder as the warlock grew closer to the cathedral. We can conclude that our music design includes attaching separate musical layers to the distance of certain objects. We will therefore attach our disconcerting, high string texture to the cathedral so that the closer the player becomes, the louder it grows. Let's also design the music such that the low string atonal texture and female choir will enter only once the player has entered the grounds of the cathedral.

So far, we know that as long as the player is wandering around with a fair distance between the character and the cathedral, our base layer of exploration music will play, consisting of a string motor with an oboe melody. As the player approaches the cathedral, even if it is not yet visible, the high string texture will subtly begin to enter. The closer the warlock gets to the cathedral, the louder this texture will become. Once the player enters the cathedral's grounds, our low string texture and choir will enter.

It is essential to realize that at this point, we have simply added musical layers on top of the exploration music to change the quality of the cue from exploratory to somewhat scary and tense, thereby changing the context of the original layer. The technique of adding or removing layers of a single cue is called *vertical remixing*, or *vertical layering*. Vertical layering is a fundamental and common game scoring technique. It is best described as "layering" (rather than remixing) because the same track plays while different layers are added or subtracted.

In order to use this technique, you must bounce out multiple stems, or layers of the music. The content of each layer depends on what is needed for each particular cue. In the case of our example, we are scoring a tense and foreboding scene that requires some arrhythmic, atonal musical layers to raise the tension as the player grows closer to the danger. Note that Layer 1, which provides the light, explorational melody, fades out when the player enters the cathedral grounds, leaving space for Layer 3 to build fear and tension, as shown in Figure 2.3.

Layers are generally broken down into groupings of instruments, each serving a particular purpose. We heard a vertical layering process occur previously when we listened to our first companion site example, generously provided by composer Alba S. Torremocha.

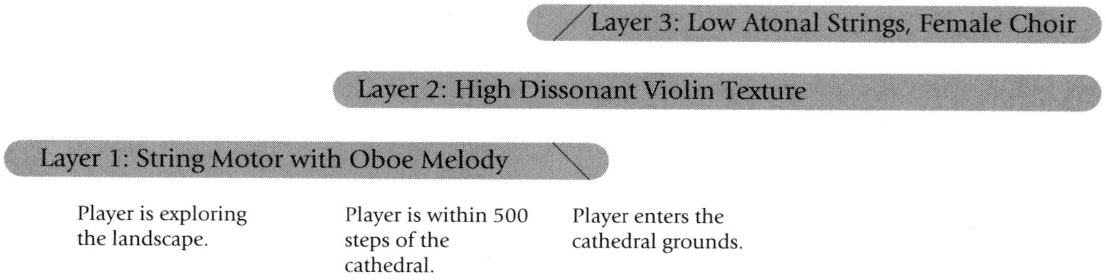

Layer 3: Low Atonal Strings, Female Choir

Layer 2: High Dissonant Violin Texture

Layer 1: String Motor with Oboe Melody

Player is exploring the landscape.　　Player is within 500 steps of the cathedral.　　Player enters the cathedral grounds.

Figure 2.3.
An example of basic vertical layering used to transform a cue from relaxing to tense as the player grows closer to danger.

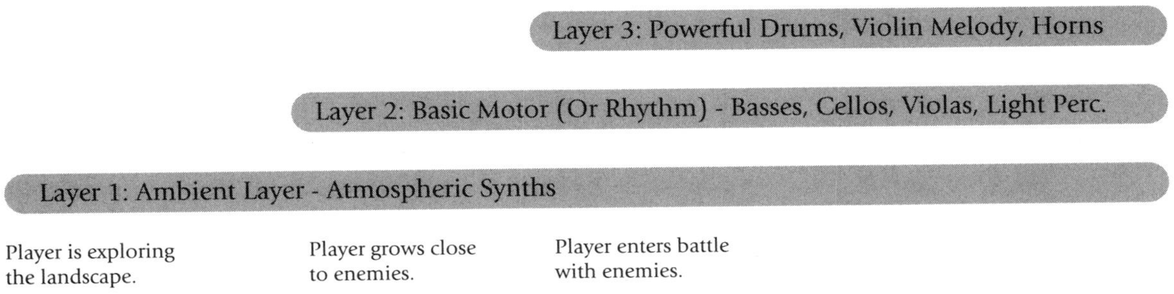

Layer 3: Powerful Drums, Violin Melody, Horns

Layer 2: Basic Motor (Or Rhythm) - Basses, Cellos, Violas, Light Perc.

Layer 1: Ambient Layer - Atmospheric Synths

Player is exploring the landscape.   Player grows close to enemies.   Player enters battle with enemies.

Figure 2.4.
Vertical layering using different instrument groupings based on their functions in the score.

Before listening to it again, take a few moments now to listen to each individual stem before hearing them all together one more time. ▶(2.11–2.13)

Red Dead Redemption (2010)[9] is another excellent example of a game that uses a complex vertical layering system. The entire score was recorded so that stems could be stacked on top of each other interchangeably to remain coherent with the action on screen. For example, if one jumps onto a horse, a bass line begins playing. As the action elevates and a chase ensues, a timpani begins rolling along with the bass line.

Rather than breaking up the groups based on instrument family alone, the example in Figure 2.4 breaks up the layers depending on their function in regard to a hypothetical gameplay. In this example, the player transitions between exploration, stealth, and combat states. Synths are used simply as an ambient layer. A basic rhythmic motive is introduced using the low strings and light percussion to indicate an increase in intensity. Finally, the third layer consists of low drums, horns, and the melody of the piece played in the violins.

Because vertical layering uses separated layers, it affords us the opportunity to use player input to affect each layer individually. We can therefore use different control inputs from the game to create musical changes. In this example, we could try using a few control inputs to give the music extra life. If the player is exploring a cave, Layer 1 could have a low-pass filter removing all of its high-frequency content. The proximity to enemies might affect the volume of Layer 2 such that the closer the player gets to those enemies, the louder it becomes, signifying an increase in danger. Perhaps whenever the player character takes a strong blow from an enemy, a washy reverb could be applied to Layer 3, serving as a musical symbol for the player character temporarily having blurry vision. Of course, these are just a few possibilities. In the end, it's up to you to determine how to use vertical remixing for your game music.

Conclusion

Most modern games use a combination of both horizontal resequencing and vertical layering. This allows a score to have a fairly complex structure that adapts to the game in multiple ways. While some moments might be best served by the addition or subtraction of a musical layer, thereby being well-suited to vertical layering, others require a complete musical shift, meaning horizontal resequencing is a preferable method for switching between musical tracks. Looping is, of course, a central part of both techniques, while transitions and transition segments are particularly key for horizontal resequencing.

With an understanding of these common game scoring techniques, you are now ready to begin the process of creating music designs for games and writing music to fit into their music systems. If you're inspired to do so, it is recommended that you dive directly into the process, experimenting as you continue to go deeper into game music composition and its techniques in the following chapters.

Exercises

Create a basic, one-page music design outline: *War Bringer: Reconnaissance* is a hypothetical first-person shooter. It focuses on creating the realism of what it's like fighting in World War II. The primary gameplay will be competitive, multi-player online matches. Music is needed to heighten the intensity and immersivity of the experience.

1. What function will the music serve? Will it be noticeable to the player at any point, or will it be purely background music? Why?
2. Will a single match have only one track, or will there be multiple options?
3. Will the track(s) loop? How long will the loop(s) be?
4. If there will be multiple options, how will these tracks differ from each other?
5. Will there be themes?
6. Do you feel that horizontal resequencing or vertical layering will best serve your score? Consider the idea of using a combination of the two.
7. How will your tracks transition from one to the next? What types of transitions will you use, and will you use stingers to highlight specific moments or ease transitions?
8. Once you have finished your music design, create your music in the DAW (digital audio workstation) of your choice! Try your best to create the necessary audio materials for delivery to the development team, including longer, looping audio files, layers, and short audio files for stingers. Include any short transitional files if you are using any composed transitions.

Notes

1. Pre-rendering music means that the music is compressed into a static audio file, most commonly as a .wav or .mp3. Obviously, once the audio is "bounced out," we become quite limited in what changes we can make to that file. One method for overcoming such limitations is bouncing out "stems," or smaller sections of instruments, separately from one another. This allows us to build scores in which stems are interchangeable, entering and exiting as needed in reaction to the gameplay.
2. Chance Thomas, *Composing Music for Games: The Art, Technology, and Business of Video Game Scoring* (Boca Raton, FL: CRC Press, 2017), 85.
3. Crossfading is a simple technique that involves fading out the first cue while fading in the second. This prevents any unnatural audio clicks that tend to occur from simply cutting off an audio file, while smoothly introducing the second track. In this book, it is assumed that you have prior knowledge of crossfading. However, if you require further information or detailed technical explanation of crossfading techniques, refer to the aforementioned *Composing Music For Games* by Chance Thomas.

4. Middleware is software that bridges the gap between composers and programmers. It is essentially like a DAW built specifically for integrating audio into games. Most middlewares (notably *Wwise, FMOD*, and *Elias*) make looping a seamless and adjustable process. It is for this reason, among many others, that middleware software is an excellent choice for game audio integration. We will discuss middleware in far greater detail later in the book.
5. Tim Summers, *Understanding Video Game Music* (Cambridge: Cambridge University Press, 2018), 23.
6. Thomas, *Composing Music for Games*, 66–67.
7. Michael Sweet, *Writing Interactive Music for Video Games: A Composer's Guide* (Upper Saddle River, NJ: Addison-Wesley Professional, 2014), 43.
8. Robert Nystrom, *Game Programming Patterns* (Genever Benning, 2014), https://gameprogrammingpatterns.com. Nystrom further highlights the four key elements of a finite state machine: First, "you have a fixed *set of states* that the machine can be in." This limits the amount of different possible states the game can move between to those that are part of the finite state machine. Second, "the machine can only be in *one* state at a time." This prevents the object from entering two states at once and has the potential to greatly save the programmer a headache. Third, "a sequence of *inputs* or *events* is sent to the machine." Essentially, some kind of input causes the state machine to transition to another state. Finally, "each state has a *set of transitions*, each associated with an input and pointing to a state." Nystrom's description here is excellent: "When an input comes in, if it matches a transition for the current state, the machine changes to the state that transition points to."
9. *Red Dead Redemption*, game disc, developed by Rockstar San Diego (New York: Rockstar Games, 2010).

3

What Is "Game Music"?
How Video Game Music Came to Be

Introduction

In the earliest years of gaming, examples of games with audio were few and far between. However, thanks in large part to the technological advances that had been made in microchip manufacturing by the late 1970s, more advanced games like *Space Invaders* (1978)[1] and *Asteroids* (1979)[2] could be released, heralding the concept of a continuous background musical soundtrack within a game. Though computer systems in this era were exponentially more advanced than ever before, they were still quite limited in raw computational ability and storage capacity. These limited resources meant that the timbre and arrangement of a game's soundscape were limited to whatever timbres the computer chip could reproduce itself. These limitations confined game audio creators to a small and highly idiomatic chip-based synthesis and low-fidelity PCM[3] sound palette. In other words, this early, low-bit sound was not merely an aesthetic choice, but resulted because of the scarcity of computational and reproductive resources available for any game soundtrack.[4] Over time, composers found ways to work around and within the hard limits of the technology available to them; their solutions led to particular stylistic conventions, which in turn informed many of the aesthetics of what would become known as video game music.[5]

Since then, the resources available for audio have grown exponentially, allowing composers to produce full-fledged, high-fidelity scores; yet, these scores still share many similarities with earlier, low-bit scores—characteristics unique to video game music. While limitations of technology influenced the low-bit aesthetic in terms of timbre and orchestration, the functional requirements of video games inevitably led game scores to sound unique compared to any other medium. This is because today's composers still face challenges completely unique to writing music for video games, many of which have existed even before technology grew more powerful. These challenges exist because of the one core difference between video games and other media—they are influenced in real time by the player. As we discussed in Chapter 1, the player's influence on a game means that the music must adapt. Thus, we use compositional techniques unique to game music, and whether we are writing low-bit, synth-based scores, or full-fledged orchestral scores, these functional requirements will result in a soundtrack with many qualities found only in game scores—qualities that mean we can refer to it specifically as video game music.

The Game Music Handbook. Noah Kellman, Oxford University Press (2020). © Oxford University Press. DOI: 10.1093/oso/9780190938680.001.0001.

In this chapter, we will explore the history and development of game music as a functional entity within the primary medium of video games, comparing and contrasting the roles it fulfills with those of the equivalent roles that music for film or television fulfill in those linear media. We will also discuss whether aesthetically related genres such as outrun and chiptunes (genres with a clear game music influence) constitute game music. We will analyze how the technological limitations inherent in the hardware of early gaming systems influenced specific compositional choices and, in doing so, will seek to define some broad aesthetic characteristics of game music. We will discuss how and why the set of formal and structural constraints unique to music composed for video games exists. Throughout this chapter, we will explore simple electronic musical scores, full orchestral scores, and everything in-between to determine which aesthetic characteristics (if any) unite the full spectrum of game music.

A First Look

Game music's technologically humble beginnings are a core consideration when analyzing its aesthetic tropes. This is not so for film music and its stereotypical aesthetic tropes, which originated directly from existing orchestral tropes and compositional conventions; for example, thanks to the standards set by early film composers such as Alfred Newman and Erich Wolfgang Korngold, the Romantic and Post-Romantic periods of Western art music composition became the chief body of influence for film music in the Golden Age of Hollywood, providing early film composers with a diverse but distinct sonic palette of musical devices and dramatic orchestral techniques already laden with specific emotional meaning to a majority of the moviegoing public. As sound film matured into the mid-century, the aesthetics of its music likewise began to mature, incorporating new elements from Neoclassical music and contemporary art compositions and building new sonic vocabularies for needs such as dramatic tension and horror. As television evolved concurrently, its music's aesthetics mirrored those of film music, appropriating the pre-existing signifiers and signifieds.

In contrast to the heavy borrowings of orchestral music aesthetics afforded to composers of film and television music, due to the strict hardware limitations of early gaming hardware, composers of game music had to look elsewhere to define expectations of instrumentation and sonic aesthetic for the medium. For example, since the earliest sound chips were capable of only monophonic sound synthesis, the grand orchestral-style music of the Post-Romantic era or the Golden Age of Hollywood—rich in timbral, harmonic, and dynamic variation—would be impossible to faithfully reproduce. Thus, early game music composers like Koji Kondo, Nobuo Uematsu, and Yuzo Koshiro had to look to newer, simpler styles of music from which to draw inspiration for integration into the likewise-simpler playback capabilities of early sound chips: popular music.

Regarding this borrowing, Koji Kondo has specifically noted in interviews that he frequently listens to pop, Latin music, and jazz, and that his scores for Nintendo games reflect this milieu, displaying influences ranging from jazz to Latin music, early rock and roll, and contemporary pop. Thus, the "Star Theme" from *Super Mario Bros.* (1985)[6] ▶(3.1) features a fast, Latin-influenced rhythmic groove, while the "Overworld" ▶(3.2) theme from the same game is a reimagining of a riff from Japanese Fusion group T-Square's "Sister Marian," ▶(3.3) released in 1984, placed over an energetic calypso beat. Even in more modern Kondo-scored games such as *Super Mario Sunshine* (2002),[7] the "Isle Delfino Theme" ▶(3.4) employs a guitar-based melody and accordion in approximation of the gypsy jazz of Django Reinhardt, while the horns and organ of "Ricco Harbor" ▶(3.5) strongly allude to early 1960s American rock.

Over the last few decades, these pop-influenced beats and harmonies, as arranged for performance by synthesizers and primitive samplers, have become key aesthetic signifiers of game music. In fact, though the computational and memory storage capabilities of gaming devices' music and audio playback systems are far greater today, many examples of modern game music continue to employ and hearken back to the classic chip-based sound; this movement is now commonly called *retro game music* or—albeit via a misnomer—*8-bit music*.[8] A parallel trend in music created for popular consumption that hearkens back to this style by using, for example, circuit-bent Game Boys is generally called *chiptunes*, although—confusingly—*8-bit music* is also common. Regardless of whether the music has been created for function or for popular consumption, however, these modern trends are driven by composers who seek to emulate, reproduce, or mangle the low-bit sounds of early game audio hardware.[9]

As a genre, such as in the chiptunes music of artists like Equip, this movement often intersects with the genres of vaporwave, synthwave, outrun,[10] and chillwave. Functional game music examples (i.e., examples created with the specific function of being integrated into a game) of this movement include Disasterpeace's scores for *FEZ* (2012)[11] and *Hyper Light Drifter* (2016)[12] and Lena Raine's score for *Celeste* (2018).[13] A fascinating hybrid of modern audio capabilities and classic chip-based game audio tropes can be found in the cues in the soundtrack to the retro, Nintendo Entertainment System[14]–style platforming level segments of *Super Mario Odyssey* (2018)[15] for the much more modern Nintendo Switch console by Shiho Fuiji, Koji Kondo, and Naoto Kubo.

The repetitive, and most importantly, interactive nature of video gaming—and, therefore, of game music—is certainly unique when compared with other types of consumable popular media. It is likely that, had low-fidelity, endlessly looping low-bit music emerged into the public consciousness independently from its primary functionality in gaming, it would have achieved very little in popular appeal. Yet, video games were a stimulating (and often, fun) medium in which short musical segments could be repeated endlessly, fostering not only intimate familiarity between the player and the audiovisual stimuli, but also an association in the player's mind between those musical thematic materials and "having a good time." Indeed, the continual repetition of any stimulus eventually forms a neural connection in the brain, for the mind to either ignore (such as the sound of a droning air conditioning unit) or latch onto (such as the sound of an approaching alarm klaxon). Thus, the direct connection built between that musical idea and the feeling of fulfillment the player experiences while playing that particular game allows the player to extend that feeling of fulfillment outside of the game. While listening to the score, the player is able to re-experience the pleasant sensation provided by playing the game. In his dissertation *Music Beyond Gameplay: Motivators in the Consumption of Videogame Soundtracks,* Juan Sebastián Díaz Gasca concludes that:

associations are created with music and are based mostly around the game elements of story and characters, and are born out of the audience's interaction with the game, and this personal interaction is what leaves a lasting impression in audiences. Musical associations are largely due to the thematic uses of music [. . .] repeated several times during gameplay, creating clear sonic signposts of the character, the personality of this character, and the abilities or actions that can be performed in the game.[16]

This same emotional connection built between pleasurable experiences and the music heard during those experiences can also be exploited in other ways. For example, popular music, which, when exploited in media, comes to its audience loaded with pre-established connotations (for example, see the prevalence of Creedence Clearwater Revival's[17] discography in films set during the Vietnam War), can be used to draw these associations purposefully during gameplay. This exploitation occurs often in video games, too, and we will later explore how we can use this technique when we discuss musical codes later in the text.

One early example of popular music being used in this way lies in the soundtrack to the *Michael Jackson's Moonwalker* (1989)[18] computer, arcade, and console games, which feature low-bit versions of popular Michael Jackson songs ▶(3.6) —an interesting case study which can help us more accurately refine our definition of game music. We already agree that game music can roughly be defined as music originally meant to act functionally during the act of playing a video game, whether consumed in that context or on its own. However, we also agree that preexisting music—in other words, music composed for a purpose other than acting functionally during the act of playing a video game—is *not* game music. We furthermore agree that the original Michael Jackson songs fully belong to this latter category, being originally from pop records. How, then, shall we classify the low-bit iterations of Jackson's songs in *Moonwalker*?

As we discussed earlier, a chief distinction between game music, on the one hand, and music for linear media, concert music, and recorded music, on the other, is the need for game music to function, interactively, at the player's discretion. Thus, at the formal level, the game designers had to modify each of the featured Jackson songs so that each piece could loop and transition appropriately within the context of the gameplay. Clearly, the formal modification enabling this looping is specific and unique to the requirements of the game versions of these songs; it is not used in either the original recordings or the cover arrangements by other artists. Thus, we can see that these *Moonwalker* songs of Jackson's were modified to conform specifically to the formal and structural needs that game music fulfills.

Furthermore, these songs took on the contemporaneous sonic aesthetic of game music, being recreated for use in-game with chips. In this era of gaming, the music found in game scores was generally conceived and originated directly in software, using trackers to control basic synthesizers and primitive samplers. In the case of the music for *Moonwalker*, however, the songs originally had been conceived, recorded, and released to the public using physical instruments and vocals; only later, for the game, were they completely rearranged for playback by the system hardware. Thus, despite their deviance in origin from that of typical game music, the arrangements appearing in *Moonwalker* still ultimately conformed to the chip-music aesthetic.

Considering the dramatic changes in formal structure (from linear to functional-looping) and in sonic aesthetic and performance (from physical instruments and voices to chip outputs controlled by a software tracker), we can see that each change shifts the essence of each track in *Moonwalker* away from the standard conventions of popular recorded music and toward the standards of game music. Thus, we can argue that though the underlying musical ideas are identical and the Michael Jackson branding likewise remains firm, the songs themselves metamorphose from artifacts of pop music to exemplars of game music.

In contrast, some games eschew this metamorphosis in their soundtracks, forgoing the transformation of popular music into music specifically meant for functional use within a game as game music, and instead allowing—if not encouraging—an experience not unlike playing a game while concurrently listening to music on a nearby radio. The *Grand Theft Auto* series[19] follows this paradigm in its in-car radio stations; so too does *Wipeout XL* (1996),[20] which allows the player to explicitly decide which tracks to listen to during gameplay. ▶(3.7) This music, though concurrent with the activities of gameplay and selected via an in-game menu, maintains the form and sonic character of the original tracks. The music of *Wipeout XL* does not exhibit any change from the original recordings in formal structure, in sonic aesthetic, or even in instrumentation. Therefore, we can see that music in games that follows this "non-transformative" paradigm should likewise not be classified as game music, but should instead be understood as a parallel to the "needle-drop" practice seen in the soundtracks to many films and television shows (as in the aforementioned CCR example).

We can thus craft a functional definition of game music, then, as music that:

1. Is, or is intended to be, used as the score to a video game's gameplay;
2. Was crafted or adapted specifically for functional use in that video game;
3. Provides some level of interactivity and/or nonlinearity.
4. Also, for games originally appearing on early gaming hardware, or to modern games created to specifically model or emulate the aesthetics common to games of that earlier era: relies on a low-bit sound chip-based sonic aesthetic.

Now that we have established *how* we define game music, let us examine the history of the category, and what forces and decisions shaped its formal and aesthetic conventions.

The Birth of Low-Bit Music

Our journey begins in 1972, when Atari released *Pong*,[21] the first video game to feature audio. *Pong* was conceived when Nolan Bushnell, one of the founders of Atari, had assigned an otherwise unremarkable engineering scenario as an onboarding and commercial-viability-test exercise for new hire Al Alcorn, challenging him to create a playable arcade game while using the least number of computer chips. This game would include two paddles and a ball, much like a game of table tennis. However, as the project progressed, the gameplay concept seemed increasingly worthy of being more than just a temporary engineering exercise. Upon realizing the game's potential, Bushnell asked Alcorn to also include sound in the game. ▶(3.8)

In an interview with IGN years later, Alcorn revealed that, as the project was already running over budget, he simply "went in [to Atari] that afternoon and in less than an hour poked around and found different tones that already existed in the sync generator, and gated them out and it took half a chip to do that. And [. . .] said 'there's the sound—if you don't like it you do it!' That's the way [the audio implementation] was left, so [Alcorn loves it] when people talk about how wonderful and well thought-out the sounds are."[22]

Little did Alcorn know that his sounds would have a big impact on the entire gaming industry, despite his self-admittedly slapdash audio-implementation process. Clearly, the hardware used to generate *Pong*'s audio was very limited; all of the pitches of the sound effects are in the same musical key, more likely due to the limitations of the hardware rather than a specific musically oriented choice made by Alcorn.

Regardless of his intent, Alcorn had also stumbled upon (and created the first example of) a game audio technique still widely used today: that of using a particular musical key across the soundscape of a game, whether in music or in musical sound effects, as a signifier of in-game location. Designing an entire soundscape around a single key enables a more seamless audio experience for the player, because to do so facilitates musical congruence between different music tracks as well as between the music proper and the musical sound effects. Keeping audio elements within the same key also allows transitional segments to function fluidly and interchangeably. Musical sound effects set in the same key as the music can blur the lines of diegesis, enhancing the overall gaming experience with sonic unity (e.g., the "get coin" sound of World 1-1 in *Super Mario Bros.* is a quick C-E heard against a piece of music in C major). ▶(3.9)

While *Pong* was the first game to include sound, and though the pitch of each sound effect did belong to a musical key, it was not the first game to actually feature music. Though games following *Pong* throughout the mid-1970s also featured musical sound effects and sometimes even began to feature short snippets of musical material in their title screens, most game audio experts agree that Taito/Midway's *Space Invaders*, shown in Figure 3.1, was the first game to be released with a truly consistent musical background during gameplay. *Space Invaders* thus marks the next major milestone in the history of game music, for it was the first game to marry all four defining characteristics of game music. As Karen Collins describes in her 2008 book, *Game Sound:* "In terms of nondiegetic sound, *Space Invaders* [. . .] set an important precedent for continuous music, with a descending four-tone loop ▶(3.10) [. . .] that sped up as the game progressed. Arguably, *Space Invaders* and *Asteroids* [. . .] represent the first examples of continuous music in games, depending on how one defines music."[23] As Collins mentions, the *Space Invaders* music is *nondiegetic*; despite being attached to an in-game parameter, the music exists outside of the reality of the game-world itself.

Many video games of the 1970s were modifications of the *Pong* mechanic—that is, two paddles and a ball, or, more specifically, two opposing paddle sprites[24] and one ball sprite—and *Space Invaders* was no exception. The direction of action on the screen may have been rotated 90 degrees from that of *Pong*, but the player would still control the first paddle sprite (here, a laser cannon) and fire the ball sprite (here, lasers) toward the second paddle sprite (here, shield barriers and the oncoming alien horde.[25] At the beginning of the game, each alien could only move (that is, be redrawn in a different location) once

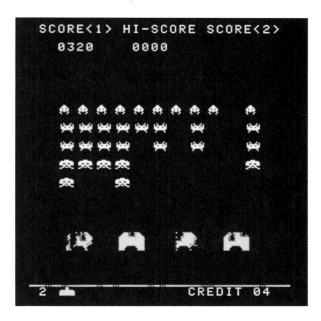

Figure 3.1.
Space Invaders, arcade, developed by Taito (Tokyo: Taito, 1978).

per second; of course, the shields did not move at all. However, as the aliens that had been shot and destroyed were then removed from the second paddle sprite's draw-buffer, the CPU[26] could compute and draw each new remaining location with exponentially increasing speed. This progression culminated in the final single alien's dizzyingly quick movements. Furthermore, as the CPU also regulated the soundtrack playback, the tempo of the music likewise increased with the removal of each alien. Tomohiro Nishikado, the designer of *Space Invaders*, states that this tandem quickening of the screen draw and of the score tempo had originally been unintentional, but he chose to keep it, as it added "more thrills to the game."[27]

So, since the tempo of the soundtrack to *Space Invaders* was directly attached to a game parameter (the number of remaining enemy aliens and shields), it was dynamically variable; the musical result depended on each gameplay. Thus, by accident, *Space Invaders* introduced the world's first interactive video game score. This score was also the first procedural score,[28] as the audio was created and performed in real time based on gameplay data.

Furthermore, the entire soundscape of *Space Invaders* was digitally synthesized, giving it the classic low-bit sound closely associated with retro game music. As we have discussed before, the realm of possibilities available for the sound aesthetic in games of this era was significantly limited by the hardware's capabilities. For example, this game was originally created on the Taito 8080 Arcade System, which employed an Intel 800 CPU and a TI[29] SN76477 sound chip. According to the manual for the SN76477, the only way to achieve two sounds at the same time (that is, to achieve polyphony) using the chip would be by the use of an external *multiplexer*, an electronic device which the manual describes as "an oscillator or frequency generator that can switch the mixer select lines at a rapid rate so that the two sounds seem to occur at the same time."[30] In other words, the SN76477 was still only truly capable of monophony; polyphony and harmony could not literally be reproduced by the chip. However, these effects could be suggested in the gestalt via rapid arpeggiations and alternations between pitches. Hard technological limitations like this prevented developers and composers in this early era of gaming from truly realizing even basic chords, much less fully realized musical scores rich in timbral, harmonic, and dynamic variation.

Continuing with our survey, the year 1982 saw perhaps the most commercially important milestone for game music: the theme to *Pac-Man* (1980)[31] ▶(3.11) became the first music written for a game to achieve mass popular recognition (here, helped largely by Buckner and Garcia's late 1981 hit, "Pac-Man Fever,"[32] ▶(3.12) which both prominently sampled the *Pac-Man* theme and had lyrics describing having a fun time at the arcade; the record was certified Gold by the RIAA[33] within a quarter).[34] Indeed, the appeal of the *Pac-Man* theme arguably marks the inception of this aesthetic of music as a genre marketable to consumers. Although the theme itself was not interactive in the way the music to *Space Invaders* was, it exhibited the sonic aesthetic characteristics unique to that era of game music. Most notably, the theme to *Pac-Man* demonstrated that a low-bit tune could be capable of endearing listeners even beyond its use in gameplay, and that—just as with music written for film or television—music written for games could also exist on popular hit records.

Pre-rendered and Orchestral Game Music

Over the course of the following decades, technological capability continued to advance quickly, allowing significantly more memory and power to be allocated to audio-specific purposes in newer gaming systems. Notably, the 8-bit Nintendo Entertainment System of 1985, originally released in Japan as the Family Computer in 1983,[35] contained a sound chip with five available sound channels:[36] four synthesizer channels (triangle wave, often used for bass; noise, often used for percussion; and two PWM[37] channels, variations of square waves often used for melody and harmony), and one channel devoted to short, low-fidelity PCM samples. The presence of this single latter channel indicated that a monumental change was on the horizon for game sound and game music: pre-rendered audio. The ability to use pre-rendered samples in a game's soundtrack meant that complex sounds, including acoustic instruments or even vocals, could now be used in game music, too.

This paradigm shift did not happen immediately, of course; "musical" sound effects from the four synthesizer channels continued to hold an important role in the soundscapes of most NES games, and composers such as Koji Kondo and Nobuo Uematsu continued to create highly reactive and interactive soundscapes using the timbres of the chip synthesizer. With the release of the next generation of consoles, however, particularly of the FM[38]-synthesis-based Yamaha YM2612 and TI SN76489 chips in the Sega Genesis (1989)[39] and the 8-voice S-SMP audio sampler chip in the Super Nintendo Entertainment System (1991),[40] an audible development began to emerge: while the sounds were still clearly synthesized and/or lo-fidelity samples,[41] each timbre began to more faithfully resemble an acoustic counterpart. With the rising voice-count and increasingly realistic timbres, game soundtracks began to exhibit more sonic detail. Thus, by the late 1980s and early 1990s, game music began to be able to more closely emulate the sounds of acoustic instruments, even whole orchestras. These technological developments allowed composers both to create more complex arrangements of their music and to begin to draw more inspiration from the same orchestral tropes that music for film and television had been using for decades.

Technology became increasingly more powerful as the 1990s progressed. For example, let us briefly consider 1994: on the SNES, the score to *Final Fantasy III*[42] ▶(3.13) by Nobuo Uematsu included very recognizable—albeit low-fidelity—orchestral instrumental timbres; *Earthbound*[43] was released, its score replete not only with references to dozens of musical genres but also with actual samples lifted from sources such as *Monty Python's Flying Circus*, *The Little Rascals*, and *Sgt. Pepper's Lonely Hearts Club Band*;[44,45] and, though merely three years after the release of the SNES, Sony's original PlayStation[46] console boasted a 16-bit DAC[47] with up to 24 channels of audio rendered at 44.1 kHz—in other words, with Red Book–compliant audio specifications, allowing not only high-resolution game audio but even audio CD[48] playback! In fact, though the PS hardware had included an SPU[49] chip, by 1996 the Nintendo 64[50] had done away with a dedicated audio chip altogether; audio was instead handled in tandem with the graphics by the main CPU.

Game music composers quickly adapted to these new technological capabilities. By the mid-1990s, it had become possible for composers to place fully realized pop-style, not to mention orchestral, scores[51] within games. With the separation afforded by multiple audio channels and increased sonic fidelity, most sound effects started to become distinctly diegetic, while most game music became distinctly nondiegetic. Part of this shift away from musical sound effects also had to do with advances in graphical capabilities; whereas sound effects had been employed previously as an experiential crutch, bolstering emotional impact in situations in which the graphical fidelity was not sufficient to achieve a certain narrative affect, now, game engines had increased enough in processing power to provide developers with the means to achieve more visual realism. As the bar for graphical capabilities of games advanced, so did the bar for the audio. Sound effects featuring distinctly musical elements would now be completely out of place in a game that was otherwise striving to replicate a real-world experience. Since gaming systems now had the power to place numerous sound effects on dedicated sound effects channels operating separately from the dedicated music channels, it also became less necessary to use musical sound design, allowing for not only more realistic sound effects, but also more separation between the respective roles of music and sound effects.

Another trend beginning in the late 1990s and continuing throughout the modern era of gaming is the use of fully pre-rendered soundtracks, no longer being performed extemporaneously during gameplay by software trackers. Composers could now take advantage of real time playback of PCM formats such as MP3 and WAV. One early game exhibiting this new trend, *Metal Gear Solid 2: Sons of Liberty* (2001)[52] for the Sony PlayStation 2,[53] boasted a hybrid-style score which combined a full orchestra with synthesized instruments. ▶(3.14) This trend of moving away from chip-based synthesized audio aesthetics toward fully realized orchestral scores led to a yet clearer separation between music and sound design.

Giants: Citizen Kabuto (2000)[54] is another early example of this trend. *Giants* is a comedic science fiction game—a genre that in the previous decades had seen a wealth of examples of chip music and chip-based musical sound effects—instead with a full orchestral score, affectatious of scores found in the equivalent filmic genre. Indeed, as

Giants contains numerous short cut-scenes, its score often functions as a linear underscore, functioning identically to scores for film and television. However, during gameplay, *Giants* uses adaptive game music–style looping techniques with its orchestral score.

The pre-rendered orchestral-style scores found in *Giants, Metal Gear Solid 2*, and other games of that era arguably foster a more "cinematic" gaming experience. However, while these soundtracks no longer sported the classic low-bit synthesized sonic aesthetic, they were still created primarily to underscore nonlinear gameplay. In other words, though the soundtracks to *Giants* and others share more functionality and sonic aesthetic with the soundtracks of linear media than, say, the soundtracks of *Pac-Man* and *Super Mario Bros.* do, we can see that they are still game music.

Defining a Genre

The use of recognizably orchestral timbres in video game soundtracks certainly redefined the sonic aesthetic of game music. Though no longer so quickly identifiable by the classic low-bit, synthesized sound, and despite embodying a very different compositional aesthetic based on an entirely different body of practice, fully orchestral game scores still sound distinctly like game music.

How?

Ultimately, the most salient feature of game music is its functionality in the nonlinear, interactive medium of gaming. Even in modern orchestral scores, interactivity and nonlinearity play a crucial role in determining the resultant musical formal structures. Whether game music was written in 1980, 1990, 2000, 2010, or 2020, form remains the key identifying factor.

Many modern games employ complex variations of both new and old implementation techniques that have important implications on how scores are constructed today. As we will discuss in Chapter 9, various combinations of horizontal and vertical sequencing may lead to highly reactive musical forms, which can adapt depending on player action. For our purposes now, let us say that the resultant musical form in any given game music is primarily driven by two aspects unique to gaming: interaction with the player, and possibilities provided by reactive technologies.

Thus, for example, though the instrumentation and arrangement of modern orchestral game music does indeed draw upon centuries of common orchestrational theory and the last century of best recording practice, in implementation those traditional techniques are reimagined and reformulated into specific methods of dividing up the orchestra into modules that can be layered, interspersed, and, most importantly, varied depending on the choices the player may make during gameplay. Each gameplay thus extemporaneously creates an individualized score unique to that player's immediate experience.

As a gameplay experience, functionally, this presents no problem. The adaptive, interactive elements of game music are its defining elements. However, for secondary exploitations—as on an OST[55] record, for example—this very facet of game music presents a conundrum: if the score is formally and structurally unique for each and every playthrough, how can it translate with integrity into a linearly fixed product, like an OST, that could be listened to outside of the game? What should that product sound like? How should it be structured?

Let us first determine what music we have in consideration for inclusion on an OST. In short, what music exists in a game? We have seen that video game scoring conventions encourage specific musical tropes, and perhaps sonic aesthetics, now considered idiomatic of game music. As we discussed in Chapter 1, the idea of scoring by location or by situation is also particularly important in games because of their nonlinear nature, leading to locational themes and atmospheric locational music, as well as event-based music unlike that found in any other medium. As early as 1986–1987, games like *Final Fantasy*[56] and *The Legend of Zelda*[57]—both for the NES—included a variety of locational themes; these included town themes, overworld themes, and dungeon themes. ▶(3.15) Around this time, situational themes began appearing in games, too; for example, combat/battle themes, stealth themes, and exploration themes.

Music in games can be more specific, sometimes appearing only once in a game. For example, boss music, a special kind of combat/battle theme, often occurs during a fight of particular narrative importance, establishing heightened gameplay intensity and helping the player better understand what is at stake in that fight. Nobuo Uematsu's "One-Winged Angel," ▶(3.16) the final boss theme from *Final Fantasy VII* (1997),[58] was famously the first track Uematsu had ever written in seven long and epic *Final Fantasy* games that included vocals; just as John Williams understood while composing "Duel of the Fates" for the final lightsaber duel of the film *Star Wars Episode I: The Phantom Menace* (1999),[59] ominous chanting in ancient languages promises a truly thrilling battle in any medium.

The functional game score in its primary usage, then, trends toward a collection of situational set pieces instead of following a clearly delineated narrative throughline. The tracks on the score serve as vignettes of the different locations and situations the player may encounter throughout gameplay.

In addition to locational cues, event-based cues, and even character-based cues, games also contain another rather unique category of functional musical elements: utility music. Because of the technical conventions and requirements inherent to video games, utility music may be found across nearly all genres and eras of video games.

For example, a game has a start, or title, screen. A DVD[60] might have a title screen, too, but the title screen music on a DVD is largely inconsequential; if it exists at all, it tends to be a short segment from or inspired by the score of the featured film or show, and is never included on its own in the related OST. In a game, however, the title screen music is often of monumental importance; this music is so important, in fact, that it often becomes iconic itself, drawing attention to itself both within and without the context of imminent gameplay. This is a relic of video games' beginnings in the arcade hall; the attract mode, the title screen sequence of an arcade game, would be loud and flashy in order to attract fresh customers and, ultimately, fresh quarters (case in point: the *Pac-Man* title theme we discussed earlier). Another, more modern, example of iconic title screen music is the haunting quasi-Gregorian theme for *Halo: Combat Evolved* (2001) ▶(3.17) by composers Martin O'Donnell and Michael Salvatori; as of this writing, seven games later, this theme is still the featured music in the title screen for *Halo 5: Guardians*. ▶(3.18)

The title screen, however, is not the only utility cue-type found in video games. However frenetic and adrenaline-inducing a game's standard mode of gameplay may be, the game will likely allow the option to press a button to pause the game and open an in-game UI[61] menu. While in this menu, the player may be able to perform actions such as peruse the inventory, look at the map, check current quests, see abilities and skills, or change gameplay settings. To do these tasks in absolute silence would break player immersion; thus, in-game menu music (often also called pause menu music) becomes yet another role for music that is unique to the structures and experiences inherent in the play of video games. Pause menu music, in contrast to the music underscoring the standard mode of gameplay, is often created to give the player a feeling of groundedness and tranquility. In his book *Understanding Video Game Music* (2016), Tim Summers describes the musical qualities of this particular category well: "Opening menu [music . . .] may establish the theme or set up a vibe of the game. Consequently, pause [menu] music may not be musically extroverted [. . .] For this reason, in-game menu music is usually in an ambient style and, in an effort to retain a link to the game world, sometimes uses a downtempo rhythmically augmented variation on a known theme [from the game]."[62]

Mini-game music is another common example of a use-case that has arisen out of the peculiarities of game mechanics. Some video games contain "mini-games" that exist within the context of the larger game world—a game within a game. In these games, the player is presented with a different screen configuration than in the game's standard mode, in which the player will play a completely different style of game than the main video game. For example, within the gameplay of the action RPG *The Witcher 3: Wild Hunt* (2015),[63] the player has the option of playing Gwent, a card-gambling game, to win in-game money. To support this change in gameplay styles (and, indeed, game genres), a mini-game is often scored with a different style of music than heard as part of the score associated with the main game's standard gameplay mode. Not that a game's standard gameplay mode is necessarily also its most prevalent; in Nintendo's *Mario Party* series,[64] for example, though each game is nominally set on a board game–style game board, the majority of the gameplay experience is based on dozens of mini-games. The mini-game music selection in the *Mario Party* series is varied enough to accommodate the vast array of possible mini-games; each music track is used for only a select few of those, adding to the uniqueness (and, occasionally, sheer adrenaline rush) of each mini-game's gameplay experience. ▶(3.19)

In addition to these common utility cues, there is loading screen music. Disc-based games, including games downloaded onto a hard drive, have a significantly longer loading time compared to ROM[65] cartridge-based games, which by comparison have loading times that are nearly instantaneous. Loading screens happen most often when loading a previous save-state of the game or when accessing a different area of the game—anything that requires a large amount of new assets to be loaded into RAM[66] for the game to function properly. While the data are being loading into the RAM, the game is in a transitional state and therefore cannot access its normal complex gameplay system. Therefore, a loading cue is usually quite simple and short, serving as a way to retain immersion, pass the player's time pleasantly, and operate as an indication to the player that the game is indeed operating as intended.

In many games, yet another kind of cue, the fanfare, may also play when a level, mini-game, or battle—especially against a boss—is won or lost. Needless to say, "win" fanfares are generally more triumphant than "lose" fanfares. These win/lose fanfares could be thought of as meta-musical sound effects, since they are generally more structurally complex than a basic tonal sound effect, are nondiegetic, and happen in response to specific event triggers.

Yet another utility use-case overlapping with musical sound effects is the short musical cues that play when specific actions within a game take place. Some games employ these short musical cues when an item has been found or a skill has been acquired, such as the "open chest" build-ups and "got-item" fanfares common in the *Legend of Zelda* series. We will explore these cases further in Chapter 6.

As these, and other, forms of utility music act in ways unique to the needs of video games, it follows that game music OSTs often include these utility cues. Arguably, the musical experience of the game is incomplete without them.

In the end, the form of a game score reflects a combination of the game's functional requirements, the composer's personality and compositional choices, and the player's personality and how it reflects his or her decision-making during gameplay. Game music structure is born through creating dynamism and reactivity as well as functionality. While any particular player's decisions won't necessarily reflect the original soundtrack in its final form (it is likely the composer who will string together a final form for the soundtrack), the composer's prior knowledge of *how* players and the game itself will interact with the score impacts this final result nonetheless. It could therefore be argued that when someone does listen to a game score, even outside of the game, all of these factors are indeed reflected in this final musical rendition: a creative collaboration between the composer, the player, and sometimes even the game engine.

Putting It All Together

Game music is a genre born out of the unique challenges posed by video game functionality, and the methods used by composers to overcome these challenges by finding ways to create timeless melodies and harmony despite the restrictions of early technology. The celebrated (or maligned) 8-bit aesthetic, including the musical devices used to overcome the limitations of the technology associated specifically with game music (like arpeggiation), continues to live on in a plethora of new games that use chiptune music. In particular, Nintendo continues to litter its games' soundscapes with musical sound effects; new games in the *Mario Party* and *Super Smash Brothers* series[67] are released with wonderfully dynamic and creative synthesized scores. Despite the raw power of modern gaming systems, many players continue to be fond of that sound.

As technology continues to advance, providing composers with faster and more powerful workflows for creating highly-detailed scores without any limitations to their instrumentation, it will be up to the next generation of composers to move the game music genre forward. With the advent of Proceduralism (real time scoring based on algorithms), as well as powerful middleware tools, scores are becoming increasingly modular and dynamic. Computers and AI continue to need heavy artistic guidance from humans, and it is we who will be tasked with discovering how game music will sound as we shape the dynamic music systems of tomorrow. The more we can be music designers, being involved from the early stages of the game-design process, the more artistic impact our scores will have.

No matter where technology takes us, video games will continue to rely on music to establish thematic material and create intrigue in start menus; to shape the sonic atmospheres of locations; to compel actions in gameplay; to interact with players; to worldbuild; to provide emotional feedback; and even to extend the connection between game and player outside the reaches of the game itself. Game music, a wonderfully popular genre across the world, is surely at the forefront of the artform of music. So as you continue your game music journey, keep up the good work, and keep breaking boundaries to advance it forward.

Exercises

1. Pick a game of your choice and write an analysis of its music within the context of game music history.
    a. When was the game released?
    b. What were the technological limitations of the music system?
    c. How did the composer overcome them?
    d. How would you classify the music's aesthetic? Is it low-bit, orchestral, or something else? What were the composer's main musical influences?
    e. How was this score created? Is it produced using real time synthesis, or was it pre-recorded?
    f. What types of cues does this game use, and how is the form of the music determined?

Notes

1. *Space Invaders*, arcade, developed by Taito (Tokyo: Taito, 1978).
2. *Asteroids*, arcade, developed by Atari (Sunnyvale, CA: Atari, 1979).
3. Pulse-code modulation.
4. For more detailed information about specific gaming hardware history and its effect on audio creation, refer to Karen Collins's text, *Game Sound: An Introduction to the History, Theory, and Practice of Video Game Music and Sound Design* (2008).
5. Also known as *game music* or *VGM*.
6. *Super Mario Bros.*, game cartridge, developed by Nintendo Creative Department (Kyoto: Nintendo, 1985).
7. *Super Mario Sunshine*, game disc, developed by Nintendo EAD (Kyoto: Nintendo, 2002).
8. Early video game sound fidelity capabilities could be anything from 1-bit (as in the Apple II) to 16-bit (as in the Super Nintendo Entertainment System), though 8-bit was indeed the capability of the sound chips in popular 1980s gaming systems like the Nintendo Entertainment System and the Commodore 64. We will, however, henceforth refer to this range of sonic fidelity possibilities with a more accurate term: *low-bit*. Likewise, as *chiptunes* refers to a genre of marketable, nonfunctional music, the term *chip music* will refer to aesthetically similar trends in the functional category of game music.
9. When audio is recorded, two measurements are taken. *Sampling rate* is the number of times per second a signal is captured. The sampling rate is directly related to capturing a signal's frequency range. The *bit depth* measures the audio signal's dynamic range. The higher the bit depth, the more times per second the signal's amplitude is sampled. The sound we are

accustomed to hearing in retro-style game music is noisy because the resolution is not high enough to capture or reproduce a high-fidelity audio signal.

10. This is a genre of electronic music that, in fact, draws its name from a Sega arcade racing game: *Out Run* (1986). However, the outrun genre usually has more in common with synthpop of the late 1980s and synthwave of the early 2010s than the music of the game itself.
11. *FEZ*, game disc, developed by Polytron Corporation (Montreal: Trapdoor, 2012).
12. *Hyper Light Drifter*, download, developed by Heart Machine (Culver City, CA: Heart Machine, 2016).
13. *Celeste*, download, developed by Matt Makes Games (Vancouver, BC: Matt Makes Games, 2018).
14. Commonly known as the *NES*.
15. *Super Mario Odyssey*, game cartridge, developed by Nintendo EPD (Kyoto: Nintendo, 2017).
16. Juan S. Díaz Gasca, "Music Beyond Gameplay: Motivators in the Consumption of Videogame Soundtracks" (Mount Gravatt, Australia: Griffith University Press, 2013), 196.
17. Commonly known as *CCR*.
18. *Michael Jackson's Moonwalker*, floppy disk, etc., developed by Emerald Software et al. (Witton, UK: U.S. Gold, 1989).
19. E.g., *Grand Theft Auto: San Andreas*, game disc, developed by Rockstar North (New York: Rockstar Games, 2004).
20. *Wipeout 2097* (*Wipeout XL* in North America), game disc, developed by Psygnosis (Liverpool, UK: Psygnosis, 1996).
21. *Pong*, arcade, developed by Atari (Sunnyvale, CA: Atari, 1972).
22. "Al Alcorn Interview," Cam Shea, IGN, published March 10, 2008, updated May 12, 2012, accessed December 16, 2019, https://www.ign.com/articles/2008/03/11/al-alcorn-interview?page=1.
23. Karen Collins, *Game Sound: An Introduction to the History and Practice of Video Game Music and Sound Design* (Cambridge, MA: MIT Press, 2008), 12.
24. "An icon in a computer game which can be manoeuvred around the screen by means of a joystick, etc.," *Collins English Dictionary—Complete & Unabridged 2012 Digital Edition* (online, 2012), s.v. "sprite," https://www.dictionary.com/browse/sprite?s=t.
25. As all of the aliens and all of the barriers were technically all screen-draws of this single second paddle sprite, and NTSC CRT monitors operate at 60 cycles per second, each cycle allowed the screen-draw of only one position of this second paddle sprite—that is to say, one of the initial group of 55 aliens and 4 shields present at the beginning of gameplay.
26. Central processing unit, the main computer chip.
27. "Nishikado-san Speaks," *Retro Gamer,* Issue 003, 2004, 35.
28. A score generated in real time (as opposed to pre-rendered by the composer) based on game data. We will discuss this concept in more detail in Chapter 10.
29. Texas Instruments.
30. Texas Instruments, *Advanced Circuits—Type SN76477 Complex Sound Generator: Bulletin no. DL-S 12612* (Texas Instruments, 1978), 6.
31. *Puck Man* (*Pac-Man* in North America), arcade, developed by Namco (Tokyo: Namco, 1980).
32. Buckner and Garcia, "Pac Man Fever," 1981, single, Columbia/CBS Records, December 1981, 7-inch.
33. Recording Industry Association of America.
34. RIAA, "Gold & Platinum" Search Portal, "Buckner & Garcia—Pac-Man Fever," accessed December 16, 2019, https://www.riaa.com/gold-platinum/.
35. Commonly known as the *Famicom* or *FC*.
36. The Famicom's architecture also allowed for game cartridges to bring additional sound chips into the audio mix, increasing potential voice count and timbral variety. However, the architecture of the North American version, the NES, precluded such additions. A classic example of the difference between the two is the differing soundtracks of *Akumajō Densetsu*

(JP, 1989) for the Famicom and *Castlevania III: Dracula's Curse* (NA, 1990) for the NES, though the games were otherwise largely identical.

37. Pulse-width modulation.
38. Frequency modulation.
39. The Japanese version, the Sega Mega Drive, was released in 1988.
40. Commonly known as the *SNES*; the Japanese version, the Super Family Computer, or *SFC*, was released in 1990.
41. The SNES had only 64 KB of RAM for audio samples. This is the equivalent amount of information it takes to store about 13.5 average English words—not counting spaces!
42. *Final Fantasy VI* (*Final Fantasy III* in North America), game cartridge, developed by Square (Tokyo: Square, 1994).
43. *Mother 2* (*Earthbound* in North America), game cartridge, developed by Ape, and Hal Laboratory (Kyoto: Nintendo, 1994).
44. The music for *Mother 2/Earthbound* filled about 30% of the entire game cartridge ROM!
45. "The Many Samples and Sound-Alikes of *Earthbound*," Chris Person, Kotaku, published June 19, 2013, accessed November 6, 2019, http://www.kotaku.com/the-many-samples-and-sound-alikes-of-earthbound-5887789.
46. Commonly known as the *PS*.
47. Digital-to-analog converter.
48. Compact disc.
49. Sound processing unit.
50. Commonly known as the *N64*.
51. Albeit, often not true orchestral recordings, but of MIDI-based sampler and rompler playback recorded as audio tracks.
52. *Metal Gear Solid 2: Sons of Liberty*, game disc, developed by Konami Computer Entertainment Japan (Tokyo: Konami, 2001).
53. Commonly known as the *PS2*.
54. *Giants: Citizen Kabuto*, game disc, developed by Planet Moon Studios (Los Angeles, CA: Interplay Entertainment, 2000).
55. Original soundtrack.
56. *Final Fantasy*, game cartridge, developed by Square (Tokyo: Square, 1987).
57. *The Legend of Zelda*, game cartridge, developed by Nintendo EAD (Kyoto: Nintendo, 1986).
58. *Final Fantasy VII*, game disc (2), developed by Square (Tokyo: Square, 1997).
59. *Star Wars: Episode I—The Phantom Menace*, directed by George Lucas (1999; Los Angeles, CA: Lucasfilm/20th Century Fox), digital.
60. Digital versatile disc.
61. User interface.
62. Tim Summers, *Understanding Video Game Music* (Cambridge: Cambridge University Press, 2018), 19.
63. *The Witcher 3: Wild Hunt*, game disc, developed by CD Projekt Red (Warsaw, Poland: CS Projekt, 2015).
64. E.g., *Mario Party 2*, game cartridge, developed by Hudson Soft (Kyoto: Nintendo, 1999).
65. Read-only memory.
66. Rapid-access memory.
67. E.g., *Super Smash Bros. Melee*, game disc, developed by Hal Laboratory (Kyoto: Nintendo, 2001).

Suggested Resources

- Karen Collins, *Game Sound: An Introduction to the History and Practice of Video Game Music and Sound Design* (Cambridge, MIT Press, 2008).
- Tim Summers, *Understanding Video Game Music* (Cambridge: Cambridge University Press, 2018).

PART II
Creating the Music Design

Developing the music design for a game is one of the most fun and creative parts of the game scoring process. It's the part during which you will communicate in-depth with the game developers about a critical question: what role will the music serve in the game? In order to best answer that question, we have the responsibility as composers of answering questions about several key concepts first: How does music help keep the player immersed in the gameplay experience (Chapter 4)? How can we use sonic cues (or sonic codes) to communicate information to the player (Chapter 5)? What are musical sound effects and what role (if any) will they play in the game (Chapter 6)? What is the relationship between the music and the other sound in the game (Chapter 7)? Finally, what type of relationship does the player have with the music (Chapter 8)? We will answer all of these questions and more in Part II!

4

Player Immersion and Music

Players do not just engage in ready-made gameplay but also actively take part in the construction of these experiences: they bring their desires, anticipations and previous experiences with them, and interpret and reflect the experience in that light.
—Laura Ermi and Frans Mäyrä, "Fundamental Components of the Gameplay Experience: Analyzing Immersion"[1]

What Is Immersion?

The Oxford Learner's Dictionaries define the verb *immerse* as "to become or make somebody completely involved in something."[2] This definition of immersion applies not only to games but to all forms of media: the more immersed in the experience of a medium the audience is, the more engaged that audience will be in that medium. Not only that, but the more the audience will become functionally unaware of the existence of a world outside of that media.

Game audio creators have a responsibility to make each gaming experience as engrossing as possible. Their choices will serve to either enhance or detract from the player's connection to the game world. There are certain factors that make game immersion unique compared to other forms of media immersion, and they stem primarily from interaction. It is important for us as composers to understand this distinction as our music and sound design is a huge part of the successful immersion of the player.

In their 2005 study, *Fundamental Components of the Gameplay Experience: Analysing Immersion*, Laura Ermi and Frans Mäyrä describe their model for analyzing immersion in games. They posit that there are three different, complementary types of immersion: *sensory, challenge-based,* and *imaginative*, which together comprise the *SCI* model of immersion analysis:[3]

1. The first dimension of gameplay experience that Ermi and Mäyrä distinguish is *sensory immersion*, which occurs when "the player becomes entirely focused on the game world and its stimuli"[4] due to the impressive and comprehensive audiovisual experience of modern gaming. This effect is nearly identical to the sensory immersion that occurs when watching a film in a theater.

2. The second dimension of gameplay experience that Ermi and Mäyrä distinguish is *challenge-based immersion*, which is ". . . the feeling of immersion that is at its most powerful when [the player] is able to achieve a satisfying balance of challenges

and abilities."⁵ In other words, challenge-based immersion occurs when the player becomes absorbed in a test of skill or ability.

3. The third and final dimension of gameplay experience that Ermi and Mäyrä distinguish is *imaginative immersion*, a "dimension of game experience in which [the player] becomes absorbed with the stories and the world, or begins to feel or identify with a game character."⁶

This analysis of the combination of sensory, challenge-based, and imaginative immersion is indeed a compelling methodology which can be used to discuss the impact that music has on the player during the gaming experience. However, in her essay "Analyzing Game Musical Immersion: The ALI Model,"⁷ Isabella van Elferen argues that the SCI model is better suited for the broader analysis of ludic immersion than for the specific immersive effects that music can have on a game and its player.⁸ The most important element that SCI analysis lacks, she argues, is a methodology with which to analyze the interaction between the player and the music. We will discuss the player and music relationship in great detail in Chapter 8, "Designing Interaction between the Player and the Music," going over different ways of designing it. However, despite these shortcomings, the SCI model is a useful tool for understanding the functionality of game music; by using SCI to first understand the patterns of general immersion in a game, observations about why certain musical choices had been made by the composer and developers become better informed.

Of the three SCI dimensions, challenge-based immersion is the dimension most unique to the medium of games, as it is a dimension grounded in interactivity. However, the soundscape in a game—similar to that of a film or a television show—is primarily integral to the dimensions of sensory immersion and imaginative immersion (although, of course, in games centered around music and/or audio-specific gameplay, it would be integral to the dimension of challenge-based immersion, too). Well-crafted audio use and implementation can serve to overwhelm sensory awareness outside of the game world and its stimuli, strengthening the player's sense of immersion into the game world. One common example of this phenomenon is often found in First-person Shooter war games, such as those of the *Call of Duty* series. During gameplay, the player is barraged with a flurry of diegetic sound effects that simulate the soundscape of real war, while simultaneously building an emotional connection to the player-character, enhancing the player's will to survive.

The opening scene of *Call of Duty: WWII* (2017), developed by Sledgehammer Games and published by Activision, takes place below decks on a ship approaching the beaches of Normandy on D-Day.⁹ A simple, yet elegant, example of a soundscape strengthening imaginative immersion (that is, the player's connection to the game's story and characters) is the rickety ship-sounds that form the environmental ambience. The scene depicts a group of friends in tight quarters surrounded by the ship's seemingly sturdy metal walls on all sides. In this moment, it may be easy for the player to forget that these characters are, in fact, in grave danger. The use of the ship ambience therefore subtly reminds the player of the realities of the game world and the fragility of the vessel when faced with weaponry meant to penetrate its walls, strengthening their awareness of the game's true perilous setting as they get to know the characters in a relatively comfortable setting. The ambience serves as a constant subliminal reminder of danger and unease; whether consciously or not, the player pictures the boat moving through the water and wonders how close the men inside are to entering the war zone. The player simultaneously grows closer to the soldier characters and feels a growing sense of fear for those soldiers' lives. ▶(4.1)

Although sound, used wisely, adds immensely to the effect of imaginative immersion by augmenting the player's absorption into the game world, we must note that, when used incorrectly, it could just as easily detract from that process. For example, imagine this same opening scene but with the cry of a distant monkey as part of the environmental ambience. This element would greatly distract the player from the conversation at hand, causing the player to wonder why there is a monkey on board the war ship, forcing the player to instead use his or her imagination to interpret the sound designer's choice. The player would be prevented from forming a close personal bond with the story and its characters (especially if the monkey never materialized—the question of its location or significance would always remain). This inappropriate sound would therefore be detracting from the process of imaginative immersion.

As the opening scene continues, the characters find themselves on an open *LCVP*[10] motorboat approaching the battle on shore. The audio here is used for both sensory immersion (barraging the player's senses with a flurry of explosion and gunfire sounds) and imaginative immersion (stimulating attention to fill in the "missing" visual elements of the story—that is, what is happening outside of the boat). The awareness of dangers that can be heard but not seen strengthens the player's sense of worry for the characters. In these ways, the audio supports game immersion, bolstering the realism of the high stakes situation. ▶(4.2)

Throughout the entire opening sequence of *Call of Duty: WWII*, as the troops land on the beach—roughly the first three minutes of gameplay—there is no music. The player is indeed already auditorily overwhelmed by the cacophony of weaponry across the beach and visually preoccupied with the countless directions from which gunfire is pinning them down. The imaginative immersive element having been firmly pre-established in the player's psyche, fomenting a deep desire to keep the protagonist alive, allows space for audiovisual sensory immersion. This is supplemented with the inclusion of challenge-based immersion to drive this hectic scene now that the cut-scene has ended and gameplay has begun. With the three dimensions working in tandem on the player's gameplay experience, the player is now equipped to properly undertake the challenge of navigating the beach without dying.

If this overwhelming and frenetic opening moment demands an absence of musical score, then the point at which the score first enters and the purpose it serves must have been likewise carefully considered. Here, the player has just successfully deployed a Bangalore Torpedo, otherwise known as a "banger," thereby destroying the barbed wire fence hemming the squad, allowing the soldiers to pass through into the trenches.

The score, written by Wilbert Roget II, strategically enters at this very moment precisely because of the resultant downward shift in sensory stimulation and overall game intensity. Up until this point, there has been a feeling of imminent danger. The player's senses have been bombarded, instilling the feeling of a complete lack of control over fate. However, after the first truly successful action of agency has been carried out, the player gains a newfound sense of accomplishment. The environment also changes: the protagonist no longer charges wildly through an open field full of gunfire, but can now maneuver through trenches, using skill and common sense to obtain cover. In other words, the battle becomes more of an even match, and the player's perception of the level of danger drops accordingly. ▶(4.3)

The emotional lull that occurs at this moment must be accounted for to avoid player boredom or confusion. The player must retain a sense of purpose and cannot be allowed to feel truly relaxed, even for a second, as this would remind the player that he or she is simply staring at a screen, interacting with a virtual scenario. This shift in intensity also negatively affects the degree of the effectiveness of the imaginative immersion; the connection to the player character is weakened if the player no longer feels the character's life is in imminent danger and must be protected at all costs. This clear shift downward in sensory and imaginative immersion would be all too noticeable—that is, without the entrance of music, which brings the effectiveness of the imaginative immersion back to the necessary level by compensating for the drama and intensity that had been lost in the shift to the new scenario.

Effective game developers, composers, and sound designers must know what the overarching goal is for the game soundscape. *Is the goal of the audio to immerse the player by creating an audio-related challenge? Is it to enhance emotion in order to make the player feel more connected to the story? Is it to contribute to the worldbuilding process? Or is it to accomplish two or more of these goals simultaneously?* These are among the questions the audio team must discuss with the developers before creating a single sound. The answers will inform decisions regarding what kinds of music-to-sound design relationships are necessary to craft, and what types of player-to-sound relationships are expected to be used. The answers will also inform what types of musical elements to use in the soundscape—for example, will there be musical sound effects, and if so, what types will they be? What codes will they reference? What functions will they cover? We will discuss musical sound effects in more detail in Chapter 6, "Exploring the Use of Musical Sound Effects in Games."

Overall, the nature of the roles that music and sound design play in ludic immersion is determined largely by the way the player interacts, and is meant to interact, with them. The specifics of these interactions are dependent on the requirements of each game and the intended player experience. Each game might also contain various types of musical and sonic interactions; these interactions are key to the effectiveness of the roles that audio plays in ludic immersion.

Conclusion

Immersion is an inherent and integral part of any gaming experience, and the decisions the audio team makes play a huge role in its degree of success. By the application of the SCI model of immersion analysis, a game's effectiveness in the three dimensions of sensory immersion, challenge-based immersion, and imaginative immersion can be explored and better understood. Though the SCI model is successful in the analysis of overall gameplay immersion, it does lack a specific analytical tool to account for the player's interaction with the music. Thus, by exploring the various possibilities of sound-to-player relationships available, and viewing the shapes and forms of the dimensions as determined by the SCI model, effective audio teams can create elegant and effective interactional plans for their soundscapes. Ultimately, these different categorizations and analytical tools are no more than tools of empowerment, allowing audio creators to think critically about how their contributions affect the player at a conceptual or psychological level, in order to create the best soundscape possible, enhancing the immersion of the gameplay experience.

Exercises

1. Choose two of your favorite games. For each, follow steps 2–5.
2. Find an example of *sensory immersion.*
    a. Does the music play a role in this example?
    b. If yes, how does it affect the immersiveness of the example?
    c. If you were the composer, what would you change or keep the same? Why?
3. Find an example of *challenge-based immersion.*
    a. Does the music play a role in this example?
    b. If yes, how does it affect the immersiveness of the example?
    c. If you were the composer, what would you change or keep the same? Why?
4. Find an example of *imaginative immersion.*
    a. Does the music play a role in this example?
    b. If yes, how does it affect the immersiveness of the example?
    c. If you were the composer, what would you change or keep the same? Why?
5. Overall, do you feel the music was utilized effectively for strengthening immersion?
    a. If yes, why?
    b. If no, what would you change to improve the overall gaming experience?

Notes

1. Laura Ermi and Frans Mäyrä, "Fundamental Components of the Gameplay Experience: Analyzing Immersion," *Changing Views: Worlds in Play—Selected Papers of the 2005 Digital Games Research Association's Second International Conference* (British Columbia, Canada: Vancouver, June 16–20, 2005), 15–27; 2.
2. Deuter, Margaret, et al. "Oxford Learner's Dictionary." *Oxford Learner's Dictionaries,* Oxford University Press, 2020, www.oxfordlearnersdictionaries.com/us/definition/american_english/immerse.
3. Ermi and Mäyrä, "Fundamental Components of the Gameplay Experience," 1.
4. Ibid., 7.
5. Ibid., 8.
6. Ibid., 7–8.
7. Isabella van Elferen, "Analysing Game Musical Immersion: The ALI Model," *Ludomusicology: Approaches to Video Game Music* (Indonesia: Equinox Publishing, 2016), 35–36. In this article, van Elferen proposes the "ALI Model" for analyzing musical immersion specifically, consisting of *musical affect, musical literacy,* and *musical immersion: musical affect* is the emotional impact the music has on the gamer, *musical literacy* is the player's past experience with specific musical conventions, and *musical interaction* is "interaction with and through music."
8. Ibid., 33.
9. *Call of Duty: WWII*, game disc, developed by Sledgehammer Games (Santa Monica, CA: Activision, 2017).
10. Landing craft/personnel/vehicle.

5

Using Musical Codes to Enhance the Gaming Experience

Introduction: Experiences and Associations

The mind is constantly interpreting stimuli and comparing the information it gathers against its past experiences and extrapolations. To do this, it first categorizes these observations (that is, contextualizes and crystallizes the information into experiences), and then predicts a likely outcome (that is, uses the current, and similar former pieces of information to extrapolate a broader reality). Thanks to this process, we can know (for example) where to place the next foot as we go up a set of stairs or how hard to throw a ball for it to reach a specific target—even if we have never ascended those particular stairs or thrown that particular ball before! In his book *The Culture Code: An Ingenious Way to Understand Why People Around the World Live and Buy as They Do,* social scientist and psychoanalyst Clotaire Rapaille discusses a consequence of this mental *modus operandi* by exploring the psychological concept of an *imprint*, an association formed in the mind through the combination of an "experience and its accompanying emotion."[1] Different imprints have different *scopes*: there are imprints that are completely personal, belonging to only a single individual; there are imprints belonging to certain demographics (such as generation, culture, or nationality); there are imprints common to all of humanity. These latter imprints are *universal codes*. In fact, some of these universal codes are even hardwired into human physiology. For example, as our ancestors were living in the wild before they had the ability to construct protective shelter, certain sounds would trigger their hair to "stand up" as they became alerted to a potential threat. Often, loud, transient sounds could trigger this fight-or-flight response. To this day, we humans retain this response.

In cases in which a certain association is shared by a group of people of a particular demographic, such as societal background or generation, these associations are called *culture* (or *cultural*) *codes*. Unlike universal codes, most of which have been programmed into our human physiology through evolution and natural selection, culture codes are established on a much shorter timescale through daily exposure. Culture codes form when a group of people is regularly exposed to specific situations in which a particular image or sound is used to signify a certain idea; this repetition enforces the neural connection between the signifier and the signified.

The research of social scientists like Rapaille and others has demonstrated that imprints have a major impact on how humans relate to ideas, objects, and even each other. For this reason, codes are an incredibly useful tool in the world of marketing: when a company needs to quickly establish its values in the consumer's mind, it uses the imagery or sounds that reflect the *codes* with which the company wishes to be associated. For an example of how companies purposefully exploit culture codes to achieve specific goals in their markets, let us examine Rapaille's involvement with the marketing campaigns for American automobile manufacturer Jeep.

Through a study held in the United States, Rapaille concluded that Americans associate the Jeep brand with the idea of "horse"—that is to say, in the minds of most Americans, the Jeep is the "'horse' of automobiles"; a rugged, all-terrain vehicle, strong and reliable, without need for the amenities found in a luxury SUV.[2] After learning this information, Jeep refocused its marketing direction toward strategically connecting and reinforcing the "Jeep-as-horse" culture code with consumers, even going so far as to return its headlights back to the iconic circular shape to make them more reminiscent of eyes. Their capitalizing on the preexisting cultural code led to a sizable rise in Jeep sales, because Jeep was able to better appeal to their target market.

In the same way that Rapaille used the code of "horse" to empower Jeep's branding, game music composers can use *sonic codes* to affect the perception of the player. We do this by strategically attaching musical building blocks, which carry specific extramusical associations, to our scores and our musical sound effects. By applying these sonic codes, which through auditory shorthand strengthen the overall efficacy of the soundtrack, we can better influence the player's gameplay experiences and extrapolations.

Sonic codes, then, are musical devices or compositional techniques—melodies, rhythms, timbres, harmonies, or even harmonic progressions—that allude to or evoke particular associations for the listener. Of all the categories of auditory stimuli, music has a particular capacity for being associated in the human mind with particular ideas. This is most likely due to the intrinsic link that exists between music and memory, as author Daniel Levitin affirms in his book *This Is Your Brain on Music:* "memory affects the music-listening experience so profoundly that it would not be hyperbole to say that without memory there would be no music."[3] Composers use this link to their advantage.

The self-referential use of musical and sonic tropes is not a new concept. Composers (and, eventually, marketers) have been using familiar sounds for millennia to strategically build emotional connections in and with the listener. In the practice of audio branding, for example, a marketing team commissions a composer/sound designer to develop a sonic signature with which to represent a brand. This task requires the composer to have an understanding of that brand's aesthetic and the ability to translate that aesthetic into audio. In contrast, the practice of audio UX[4] design focuses on using sound to help form consumer perception of a brand—for example, the sound of a car horn in a luxury Cadillac is much more harmonically "impressive" than the sound of a car horn in a respectable Buick, though the car models are largely identical, as both brands are owned and manufactured by General Motors. Whether by sonic branding or by audio UX, the sound becomes associated with the cultural concept of the brand.

In the analysis of film music, a key area of study is the use of musical codes with preexisting associations used to purposefully influence the audience experience. In his book

Hollywood Harmony: Musical Wonder and the Sound of the Cinema, author Frank Lehman refers to these uses of codes as *style topics:*

> Like timbre, texture, tempo, and all other musical parameters active in film music, *pitch design*—the stuff of chords, scales, progressions, and keys—has the ability to structure filmic expectations. Pitch design is especially good at conveying *style topics*. These are particles of culturally encoded signification, defined by intrinsic musical characteristics and capable of enforcing or reinforcing certain meanings.[5]

While sonic codes in film music elicit emotions, thereby guiding the audience experience, sonic codes in game music have a greater potential, as they have the ability to directly influence a player's actions, thanks to the interactive nature of the medium. Composers of music for games therefore must have a deep understanding of sonic culture codes in order to translate imagery into sound when creating a sonic aesthetic for their games, using them with purpose to guide the player in the right direction. Knowledge about why certain sounds carry specific associations to certain groups of people, how and why those associations have been formed, and how to best exploit those associations is therefore paramount to successful game music composition. As both universal codes and cultural codes can have many practical applications in games, game composers' attempts to engage with the player are best served by strategic use of those codes in their soundtracks.

Thus, in this chapter we will focus specifically on sonic and musical codes, including how those codes have been used in games and other media in the past and how to best utilize them in future game projects. We will explore the various elements of sound and music that evoke connotations in the mind, and how manipulation of these elements can be used to forge new mental associations. Then, we will delve into the genres of horror and period drama, taking a practical look at how games and other media in these genres have used this technique. Finally, we will take what we have learned and experienced and extrapolate some more universal applications.

Universal Sonic Codes

Universal sonic codes, being a subset of universal codes, are shared by all humans, regardless of cultural upbringing. This is because certain sounds with certain distinct sonic attributes have the capability to elicit particular instinctive responses from any human. For example, a sudden loud, harsh sound, regardless of its actual content or source, is likely to result in a reaction of surprise.

Music[6] causes the brain to release the neurotransmitter dopamine, often referred to as the "feel good" chemical because of its role in the body's natural reward system. Preliminary scientific studies of the relationship between music and the brain have demonstrated a distinct connection between the auditory and limbic[7] systems; in other words, music seems to elicit emotional response; this is certainly true anecdotally. It therefore stands to reason that, for example, the practice of using musical elements in sound effects is likely effective across *all* cultures, particularly in examples wherein those musical sound effects are used to reward the player for performing a positive action. No matter the demographic background of the player, the presence of a musical sound chemically causes a person to "feel better."⁸ Conversely, harsh sounds (like the proverbial "nails scraping a chalkboard") usually cause or exacerbate *misophonia*, a hypersensitivity to a certain sound or sounds.

Cultural Sonic Codes

Every person in the world has a relationship with music, formed first in the womb and modified over time through experiences. Two people growing up in different cultures might have very different concepts of what music is, what things sound good, and even what should or should not be expected in the context of a given piece of music. In his book, Levitin explores *schemas*, the patterns of data we humans detect and group together when forming a concept; this categorizational behavior allows us to comprehend situations and apply our knowledge of those situations when similar situations arise. He concludes that "[w]e have musical schemas, too, [which] begin forming in the womb and are elaborated, amended, and otherwise informed every time we listen to music. Our musical schema for Western music includes implicit knowledge of the scales that are normally used."⁹ This, then, is why culture codes are specific to certain demographics: in any given culture, those inhabiting the culture are repeatedly exposed to the same specific codes through a combination of everyday experience, popular culture, and media.

Therefore, any given musical code may affect different groups of people in different ways. According to Levitin, "[t]he principal schemas we develop include a vocabulary of genres and styles, as well as of eras, . . . rhythms, chord progressions, phrase structure . . . , how long a song is, and what notes typically follow what."¹⁰ Knowing this, some game developers elect to have different versions of their scores crafted for each region of release. As Tim Summers notes in *Understanding Video Game Music*, "[t]he same title [. . .] may be sonically tailored for each territory in ways beyond simply translating the game. In the case of *Gran Turismo* (1997), the game include[d] very different music in the Japanese version [than] that heard in the edition [released in] the rest of the world."¹¹

Given that many early video games originated in Japan, this distinction is important to note, as it may reveal the likely origins of some musical codes unique to video game music. In particular, the catalog of Nintendo, a Japanese gaming company, forms a crucial—and sizable—portion of video game history, and thus early Nintendo games played key roles in establishing the prevalence of certain musical codes in video games. As we have previously explored, Koji Kondo's seminal scores for game series like *Super Mario Bros., Donkey Kong, The Legend of Zelda,* and many more, owe much of their musical language to pop music genres. The near-universal appeal of these games and their scores were essential to both the popularization of chip music as a marketable genre and the codes that consumers expect to be used in musical sound effects for video games, then and now.

While it is true that different codes have been used to elicit responses depending on the cultural region, it is also important to note that video games themselves offer their own "culture"—one populated by the people around the world who play them and understand the many codes that are unique to games. As the world becomes ever more globalized, game developers from around the world are exposed to numerous games (not to mention music) from cultures geographically outside their own. This intermingling and cross-exchange has fostered an increasingly greater homogeneity of codes. Over time, signifiers that may have once served for only locally relevant references become built into the common aesthetics of games and game music, regardless of the origin of a game's development team (including its composer). In this book, therefore, the efficacy of cultural gaming codes will be treated as having a certain degree of ubiquity to any avid gamer, with the understanding that these codes might be more or less relevant, depending on the delta between the geographic cultures of the game development team and the player.

That said, there are certain tropes in music that do not necessarily translate or export universally to the understanding of players raised in other musical practices and traditions. For example, in the Western traditions, certain emotions are often associated with certain musical codes (like the minor modes' association with "sadness")—but this relationship does not exist in the framework of other traditions, like those of the Indian subcontinent. Despite the broad global culture of "gamers," different sonic codes can still have very different meanings to players raised in different geographical cultures. Though different cultures do differ in their bodies of associated cultural imprints, in this chapter, we will focus our discussion on the musical culture codes common to video games.

Sonic Codes in Games: A Primer

Universal Sonic Codes: The Building Blocks of Sound Perception

Most examples of game music and musical sound effects contain elements of universal sonic codes that influence the vast majority of players in the same ways. This universality allows game composers to accurately predict the responses players will have to the audio within the game, regardless of the respective cultural backgrounds of each player. In other words, by tapping into these universal musical codes, game composers are able to elicit similar responses to the audio from the vast majority of people. Common universal musical codes include the following:

- *Timbre* (tonal quality): dull/bright, warm/cold, round (full)/thin
  - Brash sounds elicit a primitive sense of alert fear: the "fight-or-flight" response. Sounds in this category have found particularly rich implementation in scoring the horror genre: for example, Jason Graves, well-known for his work composing scores for horror games like the *Dead Space* series, uses a plethora of such instrumental effects as stingers. These stingers surprise the player into action, signaling when the player, in the context of the gameplay, must run for his or her life. ▶(5.1)
  - On the other hand, round, full sounds tend to be comforting and relaxing. For his score to *Mini Metro* (2014), Disasterpeace specifically intended to contrast the stressful gameplay by creating a relaxing sonic experience. The soundscape is composed of instrumental timbres built with simple sine waves. There is also an emphasis on low-frequency content, creating a deep sense of foundation and balance.[12] ▶(5.2)

- *Amplitude* (amount of sound): soft/loud
    - Quiet, soft sounds are subtler, and therefore are more easily accepted as safely belonging to the soundscape by the player. Before the Industrial Revolution, truly loud sounds were few and far between; a typical pre-industrial soundscape would have largely consisted of natural ambience: birds chirping, leaves rustling, and wind blowing.
    - Loud sounds generally signal a reason to feel alert, tense, or afraid. On the most basic level, the louder something is, the closer it is perceived to be. Also, for most of human history, loud sounds were rarer and tended to signal danger or urgency of action by interrupting the otherwise tranquil soundscape. Some examples of such sounds would have been sounds from predators, like roars or growls, or even the sound of thunder, earthquakes, and other extreme natural events.
- *Tempo* (speed of sonic events): slow/fast
    - The animal mind constantly observes and categorizes sensory information or stimuli, including music. The mind can only accurately observe, categorize, and memorize so much information at a time, similar to the processing bottleneck of a computer's process buffer. Feeding the mind new information slowly (i.e. at a slow tempo) allows it to relax and process the information without "overloading." In contrast, feeding it information quickly can have the opposite effect, causing greater stress on the mind as it attempts to input a lot of information at a fast pace. Thus, a slow, pulsating sound or rhythm will naturally enhance a feeling of calm. Refer back to our *Mini Metro* for a good example of this phenomenon as well.
    - On the opposite side of the spectrum, a rapid succession of sounds, sonic events, or notes fosters a state of heightened alertness or anxiety; when the rate of information coming into the mind surpasses its "buffer," this can cause a feeling of stress. ▶(5.3)

Cultural Sonic Codes and Their Uses

By inserting culturally specific sonic codes into video game music and musical sound effects, composers can influence the player in many ways. Sometimes, these codes may be used to encourage or discourage the player to behave in a certain way. Other times, codes can simply enhance the overall gameplay experience by, for example, underscoring something funny or awkward. In any case, the game composer should be aware of how the seemingly simple choice of genre—of style, of sonic aesthetic—can greatly affect the impression of a game and its game world.[13]

This section will take us through some different cultural sonic codes found in games. Examining a variety of musical devices, we will explore how these codes can be effectively attached to our soundscapes.

Culture Codes Attached to Genre and Time Period

Generally, each genre or style of music carries pre-established culture codes. Over time, an entire genre can take on a new meaning to a certain group of people. Generally, this is because it had subsequently been attached repeatedly to a certain idea or aesthetic in popular media. We will explore this in more depth later in the chapter, but for now let us mention a few examples.

The stylistic tendencies of cool jazz, once merely a subgenre of jazz, have now become stereotypically idiomatic of the noir film genre, a genre characterized by tropes like low lighting, murder, mystery, loneliness, a "private-eye/gumshoe" character (probably also in a trench coat), VO[14] narration, and the tragedy of normal people getting mixed up in a difficult situation. Though there is very little direct reason for the sonic aesthetic of this subgenre of jazz to necessarily have been matched or attached to this genre of film, other than perhaps the relative popularities of both genres in the 1940s, early noir works like *Double Indemnity* (1944) and *Laura* (1944) firmly established the popular association. Even noir-inspired works like the bleak cyberpunk classic *Bladerunner* (1982)—with the trappings of science fiction all throughout the premise, dialogue, and visuals, and a synthesizer-heavy soundtrack to boot—still incorporated a smoky, lonely saxophone and harmonically complex string-like pads in its "Love Theme." Thus, unsurprisingly, this culture code is often accessed even in today's films and games, for example, the *L.A. Noire* (2011) main theme by Andrew Hale, which features a classic jazz ensemble playing a slow, minor-based melody, with the addition of vibraphone adding to the "cool jazz" association. ▶(5.4)

The codes attached to classical music have been carried into games quite often as well. Similarly to jazz, classical music was once the popular music of its day. In the late eighteenth century, the works of composers like Haydn and Mozart were heard across the churches, parlors, and concert halls of Austria. Though sponsored by the nobility, the music was by and large for the enjoyment of all the people, being sonically rooted in musical traditions which themselves evolved from the folk music of the Germanic peoples; indeed, Mozart once even wrote a theme and variations on "Ah! Vous dirai-je, maman," now used as the melody to "Twinkle, Twinkle Little Star," among others. Today, however, classical music—including the music of Mozart, a man who once for a party wrote a three-part canon called, in English, "Lick Me in the Arse"—has evolved in the popular mind to represent sophisticated music for a "blue-haired," upper-class audience. This is likely due to the state of classical music performance today, wherein it is rarely performed outside of grand concert halls where admittance is contingent on procuring an expensive ticket. Because the symphony orchestras occupying these halls also rely heavily on tax-deductible donations for their annual funding, a culture of affluent donors surrounding classical music has arisen, because the patrons who have that amount of money to donate get the season tickets, box seats, and other hall amenities. The more "colorful" pieces, like Mozart's party canon, are never programmed. Thus, to many consumers today, codes associated with classical music include affluence, dignity, calm, and an upper-class lifestyle, but do not include folksiness, popular appeal, or ease of accessibility. Therefore, when a composer wants to portray a certain level of order, organization, rigidity, or affluence, the sonic aesthetics of classical music are often prominently evoked.

As we discussed in Chapter 3, the technological element of hardware limitations led the earliest video game composers to choose simpler genres of music from which to draw their cultural codes in the game music they wrote. The use and recontextualization of primarily popular genres, instead of "refined" genres like classical music, gave early video game music a low position of esteem to contemporaneous critics. Despite this, game music continued to evolve and, as the capabilities of technology progressed, to incorporate a greater variety of aesthetic influences. In fact, the solutions that composers developed to work around and against the early hardware limitations created some sonic aesthetic tropes unique to video games, even after technology's progression made these tropes—rapid arpeggiations with a monophonic voice to approximate polyphony and harmony, an emphasis on the use of rhythmic delays, and other "filling" techniques—unnecessary, if not functionally obsolete. However, as game developers and composers subsumed preestablished genre codes from films and other popular media, expanding them and creatively re-employing them to maximize the performance capabilities of the lower-fidelity contemporary gaming hardware, the classic tropes of the earlier years were not discarded.

Sonic Codes Attached to Instruments

Beyond the aesthetics of genre or style, the very instruments used in a score can have a deep impact on the player's impression of the subtext and context of the game world. Even if the score is nondiegetic, the use of instruments that could not possibly exist in the diegesis of the game world could suggest musical codes to the player which feel out of place. This cognitive dissonance could be done intentionally, to great effect; for example, *Horizon Zero Dawn* (2017), a game with a setting built on an amalgamation of science fiction robotic dinosaurs and early tribal human civilization, uses a blend of acoustic and electronic instruments. However, imagine a less tasteful dissonance: had, say, Marcin Przybyłowicz used a variety of synthesizers in his score to *The Witcher 3: Wild Hunt,* the player might have been wondering when the (nonexistent) science fiction elements of the story would finally surface.

The use of specific instruments, timbres, and sonorities to convey a commonly accepted meaning—a code—is a widely used device in media music composition, including game music composition; many of these uses are to convey the setting of a location. For example, the score to the level "Frappe Snowland" ▶(5.5) of *Mario Kart 64* (1996) uses sleigh bells, an instrument closely associated with winter and snow by Western cultures because of its wide usage in Christmas music, via the traditional use of similar bells on snow sleds and the draft animals pulling them. In that same game, the level "Koopa Troopa Beach" ▶(5.6) takes place on a stereotypical tropical island, replete with palm trees and eroded rock formations; following, its score features a steel drum, an instrument borne out of and largely associated with Caribbean island music.

Another common example of an instrumental association is that of the choir. The Church has employed choral singing as part of its tradition since its inception millennia ago, and even now, music for unaccompanied choir is *a cappella:* in the style of the chapel. Choral music therefore brings a connotation of religion and, in particular, liturgical Christianity. However, the human voice is also the most natural musical instrument, being an instrument inherent to the physical structures of the human body. While one voice alone often can convey an intimate closeness, a group of voices singing together—especially in a highly reverberant chamber—may produce a certain state of naturalistic spirituality. In other words, though the association that choir music has with religion has been strengthened in the West by choral music's prolific use in liturgical Christian practices, the instrumental timbre of a chorus of voices also carries an innate naturalistic connection, making it a universal code for spirituality. This double-coding is why the choir is often employed for Christmas music (see, e.g., John Williams's film score to *Home Alone* [1990]). Of course, Christmas itself is a syncretic holiday of both pagan and Catholic origin.

Let us now turn to the conceptual antithesis of *a cappella* music: electronic music. Synthesizers and electronic manipulations tend to be the instrumental timbres of choice for (especially dystopian) science fiction films; in this vein, scores from the "electronic tonalities" created by Bebe and Louis Barron for *Forbidden Planet* (1956) through Vangelis's score to *Blade Runner* and Daft Punk's score to *Tron: Legacy* (2010) all rely heavily on these synthetic timbres.[15]

The codes we have just discussed, as well as numerous others, have been used for decades in the worlds of film and television music. However, the aesthetically eclectic and often differing visual styles of video games provide game composers with new opportunities to utilize and combine these codes in fresh ways in order to elicit fresh responses from the player.

Musical Devices for Creating and Emphasizing Codes

Beyond the choices of sonic aesthetic and instrumentation, composers can use musical devices to further integrate codes into their game scores.

Melody

Melody is arguably the most recognizable element of any musical composition—in fact, along with the lyric, melody is the only musical element of a song protected under US copyright law. A melody can be performed alone, without any chords, orchestration, or harmony behind it, and yet will be instantly recognizable. It follows that melodies are quite often directly associated with memories; see, for example, how couples have "their song," or how famous melodies are often used in pop culture to recall specific moments in history (remember our CCR example from Chapter 3). Film's wide reach as a medium has created strong links between particular ideas and particular melodies; for example, Johan Strauss II's waltz "An der schönen blauen Donau," or "The Blue Danube," has become inexorably linked to space, since its use in *2001: A Space Odyssey*'s famous sequence. This is also why the use of leitmotif—the technique of establishing a musical connection between a melody and a character, place, or idea—is particularly effective in media: this connection can be used and reused in different contexts to portray layered meanings. Howard Shore's repetition of "Concerning Hobbits" throughout his scores to the *Lord of the Rings* films, for example, is a clear example of how a specific melody associated with a specific idea can be repurposed for use in various contexts not only to remind the audience of its original associations, but also to develop further layers of meaning. The melody's original presentation in *The Fellowship of the Ring* (2001) portrays the happy, bucolic comfort of the Hobbits' home in the Shire, while its later appearances throughout the films serve to reflect the new dangers of the Hobbits' paths and the stakes of their journeys.

Harmony

When a piece of music stays in and around its key center, the growing familiarity fosters a sense of comfort and familiarity in the listener; "safe," the music avoids going to unexpected sonic places. As we discussed in Chapter 1, game composers often write town themes with this technique for the warmth and sense of "home" it brings. At the other end of the spectrum, pieces of music using nondiatonic harmonic motion will often create the opposite effect, fostering discomfort and unresolvedness, making such a practice an excellent tool with which to score situations of unease. In other words, straying from the expected harmonic path will, at the very least, be surprising to the ear, drawing attention to itself. While certain harmonic relationships between chords and the meanings of the codes therein are largely specific to a culture's body of musical traditions, the act of clearly dissonant music finally arriving at the home key still tends to bring a sense of closure and comfort to the listener, regardless of the type of music or culture to which it, or the listener, belongs.

Furthermore, certain harmonic structures and conventions are often indicative of music styles connected to specific cultures or periods of time. Often, some of these conventions become associated with specific composers belonging to those cultures and those times. For example, much of the art music written in Europe during the Post-Romantic era adheres to one or more particular sets of harmonic conventions in vogue in the late 1800s and early 1900s. The music of composers Debussy and Ravel, however, now reflects not only the sonic milieu of *fin-de-siècle* France, but also the entire movement of Impressionism (though, notably, both composers rejected the term). Beyond these broad characterizations, there are numerous cultural associations with regard to specific tonalities, chords, modes, scales, and harmonic progressions. We will touch upon these ideas again in Chapter 6 about musical sound effects, but discussion of this subject could (and does) easily fill the pages of an entire book. While we will cover more different types of examples of harmonic codes used specifically in games, further exploration of the subject outside of this text is highly encouraged.[16]

Rhythm

In addition to melody, certain rhythmic patterns and ostinati can certainly carry code associations. Latin music genres, for example, are very percussion-oriented and have a cadre of specific rhythms representative of and popularized by those genres. The genres of bossa nova and samba were both invented in Brazil, but now as a code reflect the entire continent of South America. Or, consider this: visually, the aforementioned *Mario Kart 64* level "Koopa Troopa Beach" gives no particular indication as to where in the tropics this particular beach might be located (if it is to be located on Earth at all); however, as the music features an Afro-Cuban drum groove (a rhythmic code) underpinning a steel drum (an instrumental code) on the melody, a vibe indicative of the Caribbean islands is immediately planted in the mind of the player. As many Afro-Cuban rhythms originally grew out of folk dance music traditions, there is also a broader code association of having carefree fun.

Rhythmic Density

Just as the tempo of music may be modified in order to elicit different universal responses, the rhythmic density therein can have a similar effect. The rhythmic density of a piece of music depends on the performed subdivision of the beat. For example, at a given tempo, if the beat is divided into quarter notes it will sound "slower" or "less energetic" than if it is divided into sixteenth notes. In the score to *Just Cause 4* (2018), the rhythmic density increases in tandem with the danger level of the player's situation. ▶(5.7)

Anticipation and Deception

By playing with the player's expectations and experiential extrapolations, the composer can subvert the player's expectations by using a recognizable musical code to build the anticipation of the typical consequential result and then deviating from that resolution. This is often done by following a familiar chord progression or melodic structure, and then—seemingly suddenly—veering from what is expected. Levitin points out that "[the] most important way that music differs from visual art is that [music] is manifested over time. As tones unfold sequentially, they lead us—our brains and our minds—to make predictions about what will come next. These predictions are the essential part of musical expectations."[17] Once a norm has been established, the composer can create these moments, surprising the player; this "surprise" technique can be used in different contexts to produce anything from humor (see, e.g., Haydn's *Symphony No. 94 in G Major* [1791], often called the "Surprise Symphony") to abject terror.

The famous Miles Davis quote, "there are no wrong notes in jazz, only notes in the wrong places," references this very effect. The complexity of the harmonic languages used in jazz allows for all sorts of "incorrect" musical events to occur which, when prepared properly, instead feel "earned." One example of this is the concept of using a "wrong" or "unresolved" note to create tension. Musical tension is often created when either a leading tone or a chord is held without resolution. This "non-resolving" device is a tried-and-true way of coding tension into music, building anticipation for the hoped-for resolution. Of course, the composer could then choose not to resolve for one reason or another, deviating from the expected musical course.

Point-in-Case: A Practical Look at Horror

Horror Codes

While each media genre has its respective codes, the use of sound codes and musical codes in the genre of horror warrants its own special discussion here. Because of the stylistic requirements of the genre, horror conventions have developed to be particularly idiomatic—many of them take advantage of the universally encoded visceral reactions that humans have to harsh, loud, and sudden sounds. While composers writing music for the horror film genre have the luxury of knowing ahead of time and down to the frame just when the terror will strike, and are able to therefore craft their anticipatory textures and swells to land perfectly for the optimal scare, composers writing music for the interactive medium of video games must morph the common genre conventions into new tools to effectively score a nonlinear experience. Some game music composers have proven to be particularly adept and influential in the repurposing and expansion of these idiomatic codes, such as Jason Graves.[18] Graves pushes the boundaries of horror-related instrumental timbres by constantly recording his own samples and building unique instruments to formulate his scores, including the *Dead Space* franchise and *Until Dawn* (2015).

At the most basic level, the genre of horror has long employed "unconventional" timbres with great success. Many modern horror scores further rely heavily on contemporary production and orchestration techniques for their strange and disconcerting soundscapes and ambiences. Timing is also paramount; visual scares seem to land poorly without the aid of a jarring musical stinger.[19]

Musical devices that tend to be quite common in horror scores include the following:

- *Unfamiliarity*: As we have established earlier in the chapter, a stimulus becomes "familiar" because it has been previously experienced and categorized; thus, the mind takes comfort in having already quantified an extrapolated meaning. Repetition in the context of that extrapolation cements familiarity; familiar stimuli, including sounds and timbres, come to the mind preloaded with certain code associations. It follows that subverting this pattern by instead using sounds which are novel or unknown could enhance fear and disorientation. By artificially manipulating the properties of a known instrumental timbre, or by creating a new instrument altogether, composers of music for horror have successfully unsettled many audiences.
- *Dissonance*: The *Oxford Dictionary of Music* defines dissonance as "[t]he antonym to consonance, hence a discordant sounding together of two or more notes perceived as having 'roughness' or 'tonal tension'" (Kennedy, 2013). One common dissonant interval is that of the minor second, two pitches one half-step away from each other. While consonance is comforting and feels "right," dissonance is the opposite, feeling rough, unresolved, and "wrong." In a sense, dissonance is novelty by juxtaposition, taking two comfortably familiar elements (in this case, tones or pitches) and pitting them against each other in unexpected or unpleasant ways. Because of the ease with which dissonance can establish unease, it is one of the most universal devices found in music for horror.
- *Anticipation*: Creating and maintaining anticipation—or, more accurately, dread—is an incredibly important part of the genre of horror. Part of what makes a thing scary at all is the very belief that there is reason to be scared in the first place. This is why so many works of horror foster an expectation that something bad will eventually happen. Composers for the genre can accomplish this musically by creating sonic tension via a lack of stable resolutions; as a practical example, by gradually moving the music around various key centers without ever arriving at a specific home key.
- *Surprise*: The "jump scare" is perhaps the most idiomatic device in the genre. Humans are psychologically programmed to respond to sudden unexpected movements or loud sounds with the "fight or flight" response.

Let us see an example of these musical devices in action by examining a moment occurring early in survival horror game *Dead Space* (2008), the first of the aforementioned series. The protagonist, Isaac, becomes cut off from the rest of his team by heavy glass windows as an "automatic quarantine" alarm goes off, trapping his team in another room. Ominous bass tones move in and out of the soundscape, building anticipation as Isaac is forced to watch his team get attacked by a monstrosity. They yell at Isaac to run, but still the bass tones continue. As his team continues to yell at Isaac to run, the player (as Isaac) is cued to run into a corridor. Suddenly, monsters smash down from the ceiling, and a terrifying, sweeping string glissando covered by a clustered brass section punches into the soundscape before a high-energy rhythmic action cue, stylistically reminiscent of the chase cues found in the music to many horror films, begins. Finally, arriving at an elevator, the doors shut behind him. The dissonant, high string texture slowly begins to fade away, using the code of diminution of volume to indicate that danger is subsiding. But alas! The player is surprised one more time as a monster suddenly pries the elevator doors open, accompanied by a brash stinger: one last fright before the elevator descends and the danger has finally passed—for now. ▶(5.8)

Going Deeper: Combining Codes to Create Multilayered Associations

Layering two or more code associations together may create unexpected effects in the gameplay experience of the player. In other words, stacking codes, each with pre-established meanings and connotations, creates an alchemical amalgamation; a new code with its own distinct meaning. As we briefly touched upon early in this chapter, the concept of *leitmotif* involves first training the audience to form an association between a certain narrative concept (location, action, character) and a musical element (often a melody), then recontextualizing that meaning-laden element in a different context to further facets of meaning. No exception to this practice, scores to horror films also employ code combinations, including the following:

- *Combining and modifying instrumental timbres*: Changing the timbre of a familiar sound to make it sound "wrong," the sound remains familiar enough for the player to recognize its formerly parsed and soothing nature, yet it also sounds "incorrect" somehow, disturbing the player's sense of comfort and creating apprehension and fear. One common example of this is found in the common filmic depiction of a demon-possessed person; the psychologically soothing, natural sound of the human voice is affected using sonic effects that create a dissonant (and inhuman) mix of simultaneous pitch values, metamorphosing the timbre to something familiar yet distinctly uncanny.
- *Modified melodies*: In many cases, it is not the timbre of a sound itself that is the key point of familiarity, but the melody. Notably, lullabies—paragons of innocence, calm, and predictability—are commonly used in the horror genre. One sure way of encouraging terror is to take a familiar melody intimately known by the players and changing it slightly. This is most powerful if the changes made affect its musical resolution. For example, in "Twinkle Twinkle Little Star," the first phrase of the melody is made up of two short segments (Figure 5.1). The first sets up anticipation (bars 1–2), and the second gives the resolution (bars 3–4), ultimately landing on the tonic: ▶(5.9)

Figure 5.1.
"Twinkle Twinkle, Little Star" melody (traditional French tune).

Figure 5.2.
"Twinkle Twinkle, Little Star" melody (traditional French tune), modified so that it does not resolve.

Figure 5.3.
"Twinkle Twinkle, Little Star" melody (traditional French tune), with conflicting pedal tone.

Now, let us leave the first segment as is, providing the player with a familiar melody and the expectancy of the familiar second phrase with resolution. However, let us modify the second segment, moving the melody into an unrelated key and depriving the listener of the comfort and satisfaction of the ultimate return to the tonic (Figure 5.2). Instead of creating unambiguous finality, this turn of events creates an uneasy open-endedness. ▶(5.10)

- *Combining harmonic codes*: Changing a recognizable or otherwise predictable harmonic progression to create dissonance around a familiar melody is yet another way of making music "scary"; that is, rather than changing the notes of the melody itself, as in the previous example, the harmony is changed. Often, a simple and effective (though admittedly cliché) way to accomplish this harmonic modification is to add an unsettling pedal tone beneath the melody, one that distorts the perception of the key center, making it sound dissonant and out of place (Figure 5.3). For example: ▶(5.11)

Point in Case: A Practical Look at Period Drama

Historical Codes

As we have discussed, certain stylistic music aesthetics, including harmonies and instrumentations, may be indicative of the mores of music from specific time periods in history, which in turn are used in media to reflect such periods, even if such use is not particularly founded in the realities of contemporaneous music (for example, the quartal brass fanfares of Miklós Rósza's work in the sword and sandal genre, though often diegetic, are not accurate representations of how music would have sounded in the ancient world). Despite the relative integrities of each stylistic element, these aesthetics have come to be coded, to the popular mind, with particular forms of "otherness" in time or in place. Once these codes have been established, composers may freely use them to denote time and place in further media scores, at least until such time as such signifiers become passé or offensive—see, for example, the contrast between the aesthetics of the music of the Chinese-coded Siamese cats in Disney's *Lady and the Tramp* (1955) with that of the music of the ambiguously coded cats in the live-action Disney+ *Lady and the Tramp* (2019).

Combining Historical Codes to Create Anachronism

In games, as in film, a combination of historically referential codes can make the environment feel like it is happening simultaneously now and many years ago. The FPS[20] *BioShock* series features remote cities "lost in time," cut off from the rest of the world. *BioShock: Infinite* (2013),[21] the third game in the series, is set in 1912 in a city called Columbia, floating in the sky. Much like the underwater city of Rapture, as featured in *BioShock* (2007) and *BioShock 2* (2010), Columbia is built with technology far beyond its time—the very type of utopian technology that might have been proposed by a dreamer at the height of twentieth-century American exceptionalism.[22]

The sonic landscape of Columbia reflects the particular retro-futurism that the city represents. The diegetic music exhibits stylistic codes of the early 1900s, including traditional jazz and stride piano. However, other codes exist in this world, too: soon after the player (as protagonist Booker DeWitt) first arrives in Columbia, he comes across a barbershop quartet singing atop a blimp-suspended levitating platform. Although barbershop quartets are a common early 1900s code, this particular quartet creates anachronism: it is singing the song "God Only Knows," in reality released by the Beach Boys in 1966. *BioShock Infinite* takes place in 1912, over four decades before the song should exist! This simple musical anachronism is used to convey a powerful, complex emotional message to the player; Columbia is, after all, a city shrouded in questions and mystery. Since this is the third *BioShock* game, the player already suspects that there is more to this city than the utopian paradise presented at first glance. Yet, with the knowledge that the player is aware of the disparity between fact and appearances, the barbershop quartet continues to sing joyfully, an upbeat old-timey banner behind them and two NPC[23] lovers joining DeWitt in the audience. This idyllic setting encourages the player to halt DeWitt's exploration of the city for a while and passively listen to the performance, enjoying the moment. ▶(5.12)

A moment like this in a game like *BioShock: Infinite* causes the player to stop, think, and ask questions. There is clearly something "off" about the world in this game. Why is "God Only Knows" being performed by "The Bee Sharps"? Why has the song been retitled "A Song for Columbia"? Why is a song from 1966 being performed in 1912? These questions lead to further questions: Was "God Only Knows" actually written in Columbia in 1912, or perhaps even earlier? What else is different from life in the world as we know it in this strange, anachronistic city? Perhaps Columbia's founder, Comstock, was a Walt Disney–esque figure who actually did manage to build his own City of Tomorrow. Is this utopia really everything it appears to be?

Conclusion

From a creative standpoint, it is important to remember that the sonic codes and musical associations we have referenced in these pages are only guidelines that outline the cultural conventions, stereotypes, and tropes which have been created or imagined thus far. Often, the most memorable scores are those which deftly use, combine, and juxtapose the aesthetics of past conventions to create new, surprising associations. To cover every possible musical code is beyond the scope of this book, but we now have at least a practical foundation on which to build: an awareness of the existence and exploitation of the most common musical codes and tools by which these codes can be called upon, invented, combined, and recontextualized.

Exercises

1. Pick one of your favorite games and find five examples of musical codes being used in its score.
2. Find three instances in that game that you feel would have benefited from the use of a certain musical code.
3. Apply a musical code for each of these three places by either writing it into a musical track or creating a musical element in a musical sound effect.

Notes

1. Clotaire Rapaille, *The Culture Code: An Ingenious Way to Understand Why People Around the World Live and Buy as They Do* (New York: Random House, 2006), 6.
2. Sports-utility vehicle.
3. Daniel Levitin, *This Is Your Brain on Music: The Science of a Human Obsession* (New York: Penguin Group, 2006), 162.
4. User experience.
5. Frank Lehman, *Hollywood Harmony: Musical Wonder and the Sound of Cinema* (Oxford: Oxford University Press, 2018), 8.
6. While the definition of *music* can sometimes be controversial, we will use this term throughout the text to refer to any human-organized sound that is intended to be music. In most cases, this means that sound contains either rhythm or tonal elements, but in some cases, it may also refer to atonal or arrhythmic sounds from the *avant garde*.
7. The structures in the brain that regulate emotion and behavior.
8. The documentary *Sonic Magic: The Wonder and Science of Sound* offers us a wealth of scientific knowledge about the human relationship to sound. Jerry Thompson, director, *Sonic Magic: The Wonder and Science of Sound* (CBC Television, 2016).
9. Levitin, *This Is Your Brain on Music*, 113–114.
10. Ibid., 114–115.
11. Tim Summers, *Understanding Video Game Music* (Cambridge: Cambridge University Press, 2018), 25–26.
12. Richard Gould. "The Programmed Music of 'Mini Metro'—Interview with Rich Vreeland (Disasterpeace)." *Designing Sound*, February 18, 2016, accessed May 7, 2019, designingsound.org/2016/02/18/the-programmed-music-of-mini-metro-interview-with-rich-vreeland-disasterpeace/.
13. Though it is geared primarily toward film music, the author highly recommends Frank Lehman's book *Hollywood Harmony*, which covers all sorts of genre, temporal, and stylistic use-cases.
14. Voice-over.
15. It is interesting to note that both *2001: A Space Odyssey* (1968) and *Star Wars* (1977) both eschewed this trope by using post-Romantic fully orchestral music (in the former's case, quite literally).
16. Refer to Lehman's *Hollywood Harmony* for a detailed analysis of musical codes in film and popular culture.

17. Levitin, *This Is Your Brain on Music*, 123.
18. It is curious how last names can sometimes determine these things; James Horner wrote a lot of very good brass parts!
19. Editor's note: this is why the *Paranormal Activity* franchise comes across as predictable and bland: no music!
20. First-person shooter.
21. *Bioshock Infinite*, game disc, developed by Irrational Games (2K Games, 2013).
22. For example, Walt Disney is famous not only for his theme parks, but for his dream to build a "City of Tomorrow," of which he first envisioned an "Experimental Prototype"—thus was born *EPCOT*, though the concept had evolved into a technology-centric amusement park by then.
23. Non-player character.

Exploring the Use of Musical Sound Effects in Games

Introduction: What Are Musical Sound Effects?

One of the most fascinating elements of the medium of video games is its creative use of sound effects: brief, event-triggered sounds, apart from the dialogue or music. Many game sound effects, especially in more "realistic" games, are similar to the hyper-realistic Foley sounds found in film and television; footsteps, gunshots and swishing brush tend to sound more or less as they might in the real world. However, the medium of video games often features another, distinctly idiomatic style of sound effect: the *musical sound effect*.

Musical sound effects are sound effects (and thus are brief, event-triggered sounds, apart from the dialogue or music) which nonetheless contain a distinctly musical element as part of their sonic essence. Combining sonic codes and thematic materials, musical sound effects can serve various important functions in video games: accentuating emotions, worldbuilding providing signals and clues, and other purposes that we will explore in this chapter. Like we observed in Chapter 3 regarding the early aesthetic elements of game music, the early aesthetic elements of musical sound effects were also originally born out of technological necessity; the capabilities of early sound chips did not allow composers and audio designers the ability to record real sounds and play them back in the game, providing only basic synthesis (often via the same synthesizer being used simultaneously for performance of the music score) to suggest sounds for similar effects. With this set of limitations, composers and sound designers turned to musical elements to bolster the effectiveness of their game audio, despite the inherent simplicity. This creative reimagining and repurposing of musical elements into sound effects was not a negative, though; the colorful, fictional visuals of the worlds of early games often required a complementarily fictional approach to their respective sound designs, in order to create cohesive aesthetics. Thus, in many cases, musical sound effects were not just a "make-do" substitute for more realistic sound effects but, in actuality, a creatively appropriate choice.

Since *Pong* first incorporated sound in 1972, musical sound effects have been utilized in video games to great effect, fostering interactive worlds that both communicate emotions directly to the player and react to the player's gameplay actions. As musical sound effects are often triggered directly by the player's actions in the game (as opposed to nondiegetic underscore, which generally reacts to broad changes in game state and, therefore, only indirectly responds to the player's actions), they can offer distinct and fruitful opportunities in which to utilize musical codes. Many of these musical sound effects are composed entirely of musical elements like pitch, harmony, and/or rhythm; some are composed of a combination of non-musical sound with musical elements layered in.

This chapter is broken up into three main sections, following this introduction. In the following section we will explore a variety of ways in which musical sound effects are used in games. The subsequent section will focus on the analysis of the use of musical devices in musical sound effects, building on our previous explorations of musical codes and noting how these codes are applied when used in sound effects. And the penultimate section (preceding the conclusion) will discuss the basics of two important techniques for creating musical sound effects: synthesis and sampling.

Using Musical Sound Effects

Action Enhancement and Player Feedback

The most common function of the use of musical sound effects is to provide emotional feedback to the player in order to reinforce a specific gameplay behavior. Let us explore one such example in *Mario Kart 64* (1996),[1] a game that employs a plethora of musical sound effects throughout its gameplay experience.

In *Mario Kart 64*, as in all the three-dimensional games in the *Mario Kart* series, floating boxes containing item power-ups are found throughout each race track. Driving through one of these boxes triggers an animation: the box spins quickly, then slows down; then, if there is room in the player's inventory, the player is rewarded with an item. Sonically, starting from when the item box is driven through and throughout its spinning, a C-major arpeggio mirrors (via pitch) the rotating faces of the cube until finally the game credits an item to the player's inventory. Furthermore, the use of a major tonality serves as positive reinforcement for this action, heightening the sense of achievement the player experiences when driving through a box, thus encouraging the player to do so again. ▶(6.1)

In scenarios in which the sound effects and the music are sonically and aesthetically separate—the use of musical sound effects particularly stands out.[2] Therefore, using one at the right time is an effective way to inform the player that something important has happened, or, similarly, the degree of importance of a particular event. Because musical sound effects contain one or more musical elements, they are able to convey the same emotional content that musical codes and gestures convey. Picking up a weapon, for example, often has little to no particular narrative significance in a game; however, if a weapon is an object of notable importance, picking it up can be game-changing. Using a musical sound effect in this instance informs the player that this weapon is significant, encouraging the player to treat it with care or to explore its functionality as a game-changing item. A musical sound effect used in this way becomes a concise narrative device. Many games therefore have different versions of the "get-item" sound, depending on the significance of the object: a slightly triumphant sound for finding an item of lesser importance, a moderately triumphant sound for picking up a noncritical (but perhaps very useful or rare) item, and a highly triumphant sound for picking up something absolutely essential to progressing in the game. Perhaps no other series of games exhibits this behavior more clearly or famously than *The Legend of Zelda*.

The Legend of Zelda: Link's Awakening (1993),[3] released for the original Nintendo Game Boy, is a particularly iconic game in the broadly fantastic series of *The Legend of Zelda* games. In *Link's Awakening*, the music and sound design relationship is mainly secondary; the music of the game is mostly nondiegetic, while the majority of the sound effects throughout the game are non-musical. However, *Link's Awakening* does occasionally use musical sound effects. To highlight, for example, the importance of picking up an item, the game includes two fanfares. The standard item pickup fanfare ▶(6.2) is used for small, less-significant item pickups, while the extended fanfare is used only for when Link attains the most important game items ▶(6.3) —for example, his Shield, as seen in Figure 6.1.

While the Shield is of course central to Link's arsenal, as it is his main defensive item, the player (as Link) does not have to do much in order to acquire it—in fact, it is one of the first actions allowed once the opening cut-scene ends. This Shield is incredibly important, but not particularly special or rewarding to receive; thus, its acquisition receives the extended fanfare, a triumphant sound, but one used for a number of items throughout the game.

Finding the Sword, on the other hand, triggers a one-of-a-kind musical sound effect, more special than the extended fanfare, that plays when it is picked up. ▶(6.4) In order to acquire the Sword, the player, using only the Shield, must navigate a tricky series of enemies before finally arriving at the beach area where the Sword is. This puzzle constitutes the player's first real gameplay challenge. As Link finally draws near the

Figure 6.1.
Link receives the Shield. *The Legend of Zelda: Link's Awakening*, developed by Nintendo EAD (Kyoto: Nintendo, 1993).

Creating the Music Design

Sword, he is confronted by a magical, talking owl who makes first mention of a critical part of the game's lore, as well as a narrative goal for the player to be excited about: "It is said you cannot leave the island unless you wake the Wind Fish. . . ."[4] The acquisition of the Sword, as seen in Figure 6.2, is therefore a moment ensconced in player achievement; a perfect moment for the use of a sweeping, triumphant musical gesture—in this case, a G-major progression ensues in quick descending arpeggios (Gmaj7, Amin7, Bmin7, Cmaj7, D7) (Figure 6.3).

Providing Clues

Signaling information to the player is another common use of musical sound effects. Continuing our look at *The Legend of Zelda: Link's Awakening*, let us now explore the use of musical sound effects as used in the game's dungeons.

Soon after entering the first dungeon, Link is able to retrieve the Compass. It is revealed that the Compass will make a particular sound (a musical sound effect) when the player enters a dungeon room containing a key. This musical sound effect is a short, four-note melodic fragment: G♯-A♯-B-F♯. ▶(6.5) This melodic fragment becomes an

Figure 6.2.
Link acquires the Sword.

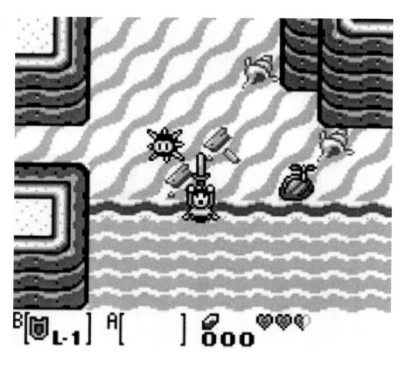

Figure 6.3.
"Get-Sword" motif from *Link's Awakening* (1993).
Music composed by Minako Hamano.

important aid while navigating the dungeon, as the player cannot proceed without finding keys, and many of the rooms in which keys are hidden do not immediately broadcast such. In other words, without the aid of this musical sound effect, the player could easily miss opportunities to retrieve numerous keys, making the gameplay experience even more frustrating than it already is, in a game already riddled with other hidden objects. Thus, in this case, this musical sound effect is critical to the gameplay, providing the player with direct clues so that important game objects are not missed.

Worldbuilding

Sonic worldbuilding is the process of using audio to supplement the aesthetics of the visual world, highlighting essential elements that reveal aspects of that world, such as how it functions, how the player interacts with it, and who (or what) inhabits its environment. As an example, *Where Shadows Slumber* (2018)[5] is a mobile 2D puzzle game in which shadows are used to create multiple versions of "reality." When audio team PHÖZ (composers Alba S. Torremocha and the author of this book, Noah Kellman) first joined the *Where Shadows Slumber* team, it was given a significant worldbuilding challenge: As *Where Shadows Slumber* was being developed for the mobile gaming market, there were limitations to what could be rendered on-screen, both in terms of the raw capabilities of the hardware and in terms of what visuals could make aesthetic sense when rendered on such small screens. PHÖZ therefore had to sonically supplement the limited potential of the visual elements in order to ensure that the overall experience of the world of *Where Shadows Slumber* would still be as immersive as possible.

The world in which *Where Shadows Slumber* takes place was designed by the developers to be reminiscent of Earth but also feel somehow "alien." Notably, the world has a "magical" quality, exhibited most clearly by its core game mechanic: the use of a lantern to control shadows in order to reveal alternate versions of the world. Following this duality of fantasy grounded in reality, PHÖZ chose to design a number of sound effects based on real-world examples, combining their known, realistic qualities with musical elements, rendering the effects simultaneously unfamiliar and alien. Thus, the player would be able to accept each sound—and therefore each soundscape—as if it were something real, yet also strange and otherworldly.

In World 0, the player is engulfed in a lush jungle (see Figure 6.4). Though there are no birds to be seen on the screen, the *Where Shadows Slumber* team wanted the ambience to live and breathe. Thus, PHÖZ created a randomized system to play flute and violin-based birdlike effects, creating a soundscape reminiscent of the presence of birds, but with a magical, "othering" quality, which was accomplished by using musical instruments instead of bird recordings to create those sounds. Indeed, the sound effects warble between pitches, much like the melodies of real birdsong; yet, their airy, cognitively-dissonant quality trends the effect toward the alien. ▶(6.6)

By using musical instruments to emulate the sounds of birds, PHÖZ accomplished worldbuilding in two ways:[6] first, by the establishment and reinforcement of the "strange-but-familiar" aesthetic of the world itself; second, by adding to the diegesis of the environment in ways beyond direct visual reference. In other words, the "birds" only exist in World 0 of *Where Shadows Slumber* because the soundscape suggests that they do.

Creating the Music Design

Figure 6.4.
Where Shadows Slumber (2018), World 0. *Where Shadows Slumber*, download, developed by Game Revenant (New York: 2018).

Without that soundscape, there would be no indication of the existence of birds. Thus, PHÖZ sonically not only reinforced the visual elements of the game world, but also created elements outright, establishing the existence in this world of creatures who are heard but not seen. This type of extra-visual worldbuilding is a powerful technique, but should be used wisely; as we covered in our discussion about immersion in Chapter 4, carelessly adding such elements could lead to an incoherent, and therefore confusing and poorly immersive, game world.

Interactivity

Continuing our look at *Where Shadows Slumber*, we can see a prime example of interactive musical sound effects in World 3—Level 1. In World 3, the player must solve his or her way out of a dark and mysterious sewer full of magical lights, which are attached to mechanical dollies. The player can interact with these lights, turning them on and off by pressing a button, as seen in Figure 6.5. PHÖZ decided to implement the audio in this section such that whenever one of these lights is turned on, a musical tone—the sound of a celesta—will sound, simultaneously adhering to the sonic aesthetic of the background music score and allowing the player to form a melody by turning various lights on in order. ▶ (6.7) Thus, the sound effects are serving as a directly interactive feedback mechanism with the player, as well as more generally functioning as a sonic element in the worldbuilding.

This interaction strengthens the "magical" qualities of the level by bridging the gap between its diegetic and nondiegetic sounds. More importantly, it further emphasizes the importance of the player's interactions with light throughout the game, adding yet another unique ability to that light.

Ambience

In some cases, the ambience of the game world is built into the music score, either completely replacing the background sounds inherent to that world or being layered in as part of them. Sound designers also often speak of the "tone" of the ambience. When designing a soundscape, they may choose a frequency center or pitch upon which the rest of the soundscape is centered, in order to enhance tonal congruency.

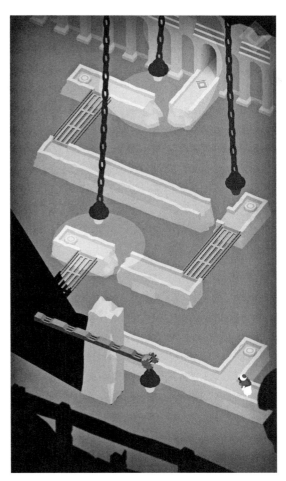

Figure 6.5.
Where Shadows Slumber (2018), World 3, Level 1. As the player presses the buttons shown in the image, different notes sound in the music, allowing the player to compose a melody.

Finally, in some cases, an otherwise non-musical diegetic sound effect becomes musical by being placed deliberately within the music score itself, typically as a rhythmic or percussive element. Take, for instance, Ryo Nagamatsu and Koji Kondo's "Cavern Theme (Going Underground)"[7] from *The Legend of Zelda: A Link between Worlds* (2013).[8] ▶(6.8) This nondiegetic theme uses the diegetic sound of water dripping in a cave in a rhythmic way; the "drop" falls on beat three of the second bar of the melodic phrases. This technique was common in older games, too. For example, *Donkey Kong Country 2: Diddy's Kong Quest* (1995)[9] also features non-musical diegetic sound effects used rhythmically in its nondiegetic score, such as in the track "Bayou Boogie" ▶(6.9) as featured on the level "Barrel Bayou";[10] in this case, the sound effects resemble the sounds of swamp animals. However, though sound effects used in this way are indeed used musically, they do not qualify as musical sound effects, as they do not themselves contain a musical element.

Designing Musical Sound Effects

Now that we have seen a number of examples of how and why musical sound effects are used, we are ready to start learning about designing our own. The first step in the process of successful musical sound effect design is determining where and when the musical sound effects will be used and for what purpose. Once this step has been achieved, you are ready to move on to the next design step: choosing the musical characteristics of each effect. The successful design of musical sound effects requires an understanding of the uses and common signifiers of musical codes. As we have learned, each sound serves a very specific purpose, whether it be enhancing action, providing feedback, signaling the player, worldbuilding, or another purpose altogether. Providing the wrong signifier in a specific situation can influence the player inaccurately, causing confusion. For example, using a musical code with a negative or discouraging connotation at the wrong moment might cause the player to avoid or altogether miss carrying out an important action; or, a sound that feels particularly positive might reinforce an action that is unnecessary or even detrimental to the player's success or enjoyment.

Therefore, once the purpose of each individual sound is determined, the effective composer should consider the various musical devices at his or her disposal with which to sonically achieve that purpose. In this section, we will discuss several musical devices and how they are used to create effective musical sound effects.

The Building Blocks of Musical Sound Effects

Ultimately, each musical sound effect found in a game is composed of a variety of elements carefully blended together to elicit the correct gameplay response from the player. Some are melodic, while others are solely harmonic. Some are based in chromaticism, while others are based in diatonicism. Still others are purely rhythmic, using percussive elements to convey their semiotic message. In navigating which codes and musical devices to use in creating effective musical sound effects for games, it helps to first be aware of the most common codes and musical devices traditionally used in the medium, as described in the following.

Melody

The use of *melody* in a musical sound effect can be exceptionally powerful in many game scenarios. As we discussed in Chapter 1, recognizable themes are often leitmotifs heard as the most prominent melody in an orchestration, forging mental associations to specific characters, places, and ideas. In like manner, the use of melody in a musical sound effect can create an emotional anchor or connection to the associated event. Each melody is itself composed of a variety of building blocks that inform and influence its coded meaning. In the West, for example, a major melody might sound happier and more positive than a minor one, which is more likely to sound sad or discouraging. Similarly, the contour of a melodic line also carries certain implications: as we will see, a generally downward shape is often deflating, while a generally upward direction is often uplifting. One example of a melodic musical sound effect using both a major tonality and an ascending contour to signal positive gain is the *Super Mario Bros.* "1-Up Mushroom" musical sound effect. ▶(6.10) This effect is composed solely of the melodic line shown in Figure 6.6.

The C-major 9 chord outlined gives the melody a positive connotation, while the ascending melodic contour simultaneously creates a feeling of achievement. This

Figure 6.6.
"1-Up" motif from *Super Mario Bros.* (1985). *Super Mario Bros.*, game cartridge, developed by Nintendo Creative Department (Kyoto: Nintendo, 1985). Music composed by Koji Kondo

economical use of positive reinforcement codes encourages the player to continue to seek out "1-Up Mushrooms" by emotionally enhancing the sense of accomplishment the player feels when acquiring an extra life.

Arpeggiation

Arpeggiation is often used for the creation of fast musical sound effects. As a musical device, it allows the composer to create quick and dynamic lines that outline a specific chord or chords. In *Duck Hunt* (1985),[11] a short musical cue, consisting of the progression Dm-G-C, plays each time the player completes each round of 10 shots. ▶(6.11) However, if the player manages to hit a duck with every single shot—a perfect round—the game rewards the player with a different musical sound effect. This effect, placed in a higher octave than the normal ending effect, is a fast-paced, ascending C-major arpeggio. ▶(6.12) Much like in the "1-Up Mushroom" example, this sound effect serves as positive reinforcement, creating a sense of triumph for the player.

Runs

A *run* is a musical effect created by performing the notes of a scale upward or downward quickly enough that they blur together—in other words, an embellishment comprised of many pitches parsed as a single directional gesture. Runs can be based upon any number of tonalities, keys, or scales (each one lending a differently coded connotation) including the following:

- *Chromatic* runs are quite common. These are fairly code-ambiguous since they do not necessarily revolve around a specific key center; therefore, these can be used in many disparate circumstances (hence their popularity). Instead of relying on a coded tonality to act as signifier, the direction of the chromatic run informs its use:
  - *Ascending chromatic runs* are often used for power-ups, sounds that reflect some sort of gain for the player. One classic example of an ascending chromatic run is the musical sound effect that plays when opening a chestful of gems in *Spyro the Dragon* (1998). ▶(6.13)
  - *Descending chromatic runs* are often used for the sound of depletion of a resource or even death (that is, the total depletion of health). This latter case is audible in *Duck Hunt,* occurring when a duck, having been successfully shot and killed by the player, plummets to the ground. ▶(6.14) The descending chromatic run in this particular case fulfills a secondary code, as a descending line is often associated with the act of "falling" itself.
- *Diatonic* runs are based upon a specific mode or scale that informs its character. For example, a run based on a major scale can sound quite optimistic, whereas a run based on a whole-tone scale can signify a more mysterious, unresolved moment.

Chords

Sometimes, even a single chord can be used as an effective musical sound effect, like that used for the explosion of a bomb in *Uurnog Uurnlimited* (2017),[12] by Nifflas Games. In this case, the sound effect is a minor 9th chord. When it plays, the root of the music also changes to fit this new chord. ▶(6.15)

Percussion

Percussive musical sound effects are those that are unpitched and/or primarily use percussive elements, whether simply via a single percussive instrument, or a repeated rhythmic figure.

- *Rhythmic*: A sound effect like this usually consists of a short rhythmic figure played by percussion instruments. In some cases, a musical code might be best conveyed with percussion due either to an association with specific percussion instruments, or a rhythmic groove. For example, in *The Legend of Zelda: Breath of the Wild* (2017),[13] the Korok sound effect is made up of three main elements and plays a short rhythmic figure. Given that Koroks are plant-like creatures, one can assume that they are made of wood. The playful rhythm compliments the playfulness of the character, while the wooden timbre of the percussion instruments also enhances the connection to trees, plants, and nature. ▶(6.16)
- *Nonrhythmic*: This type of effect is also unpitched and is based upon a percussive instrument, like in *Spyro the Dragon* (1998),[14] wherein the timbre of the sound effect attached to the magical lifts found throughout the game closely resembles that of a *mark tree*.[15] ▶(6.17)

Harmony

Combinations of chords, and the movements and positions of chords in relation to each other, have the power to convey complex ideas and emotions. Let us survey some harmonic device codes that are common to music written for games and other media:

- *Major* (including *major sixth* and *major ninth*): The major tonality is most often to convey positive feedback or to encourage a particular action. For example, take the sound effect that plays in *Breath of the Wild* when the player finds an item: this sound effect consists of a quick progression that moves from F♯ to B to F♯, and can be heard when opening up a chest to find an item.[16] ▶ (6.18)
- *Whole-tone*: Since the interval between each consecutive pitch of the whole-tone scale is exactly the same—a major second—this tonality carries a certain ambiguity and lack of tonal center. In other words, besides their "starting" notes, the "C" whole-tone scale is exactly the same as the "D" whole-tone scale (see Figure 6.7). Because there is no functional resolution inherent in the tonality, its use creates a "floating" effect, producing a certain dream-like quality. One example of an effective use of the whole-tone scale is the sound effect that plays when collecting jigsaw puzzle pieces in the puzzle platformer *Braid* (2008).[17] A bell-like timbre quickly outlines a D augmented chord, which originates from the D whole-tone scale. ▶(6.19) One of the game's core

Figure 6.7.
Note that the C whole-tone and D whole-tone scales share the same exact notes, the only difference being their starting notes (C and D, respectively).

mechanics is being able to rewind time in order to solve puzzles, while the collection of jigsaw puzzle pieces is one of the central goals. Arguably, the use of an open-ended, dream-like tonality as puzzle pieces are collected reflects the impermanent and ambiguous nature of time in the game.

- *Minor* (including *minor ninth*): The minor tonality is often used in negative contexts—perhaps most notably, in "game over" screens, as in the iconic "death" cue from the *Metal Gear Solid* series.[18] ▶(6.20)
- *Diminished*: The diminished tonality is built on minor thirds and the leading tones that connect them. A fully diminished seventh chord (1, ♭3, ♭5, ♭♭7) traditionally "begs" for tonal resolution. As such, the diminished sound has been useful in utilitarian applications like "alert" sounds, indicating to the player that she is in some kind of danger. One iconic example of such from the *Metal Gear Solid* series is the quickly-arpeggiated diminished seventh chord used to convey to the player that he has been spotted by an enemy NPC. ▶(6.21)
- *Atonal*: Atonal musical sound effects are those created by a pitched instrument in such a way that a purposeful lack of tonal center is achieved. While the whole-tone scale still retains a certain recognizable tonality (despite its being somewhat tonally ambiguous), an atonal sound effect has no key center whatsoever and is likely designed in this way purposefully. Because atonal sounds can be off-putting due to their lack of a home key, the horror genre employs atonality quite effectively. Refer back to our Chapter 5 *Dead Space* examples to see this in action.

Creating Musical Sound Effects with Synthesis and Sampling

There are two prevalent techniques used today that serve as excellent methods for creating musical sound effects: *synthesis* and *sampling*.

Synthesis is the practice of building a sound by the combination and modification of sound waves. There are many varieties of synthesis: *additive, subtractive, frequency modulation* (FM), *linear arithmetic* (LA), and *phase distortion* (PD), to name a few. The sound chip synthesizers found in many older game consoles and sound cards are most often either subtractive synthesizers or FM synthesizers.

Sampling—more specifically, *sample instrument creation*—is the art of recording a short sound and using that "sample," or a collection of such samples, to create a playable *virtual instrument*.[19]

Many of the musical sound effects we have discussed thus far originated in the classic chip-based synthesis aesthetic. As we observed in Chapter 3, this style originally came about because early gaming hardware was limited to real time chip-based performative synthesis of the game soundtrack. As we consider the practicalities of the synthesis of musical sound effects, it is important to first understand foundational synthesis theory, seeing how and why synthesizers work the way they do. First, though, let us take a closer look at the fundamental sound-origination module of any synthesizer: the *oscillator*.

What Is an Oscillator?

An *oscillator* is the analog or digital element within a synthesizer that produces one or more specific kinds of waveforms; software synthesizers, though immaterial, still have functional virtual oscillators. The waveform signals created by the oscillator or oscillators are then routed into different synthesis *modules*, sections of the synthesizer that specialize in performing certain actions on the signal, including filters, audio effects, and amplifier output. Sometimes, these modules can even be additional oscillators; for example, the *low-frequency oscillator* (LFO), which is used to create an infrasonic waveform that is to be applied to "shape" another module's output. Applying an LFO to an amplitude module will create a tremolo effect, as the signal's volume level will "slowly" (here, less than 20 times a second) modulate; applying it to a filter will cause the filter to sweep up and down at the frequency at which the LFO is set.

Subtractive Synthesis

Subtractive synthesis is the most common kind of synthesis found on classic gaming hardware. It is also perhaps the simplest kind of synthesis not only to comprehend, but also enact in hardware and software-based synthesizers. Because of these reasons, it is an ideal place to start our exploration of synthesis. Let us now examine the basic components of subtractive synthesis: *envelopes, waveforms*, and *filters*.

Envelopes

The "shape" of a sound over time is called an *envelope*.[20] An envelope consists of four basic elements: the *attack, decay, sustain,* and *release*, commonly referred to as *ADSR*. Attack, decay, and release are all parameters that denote a length of time, whereas the sustain is an amplitude level. To examine each parameter in more detail:

1. *Attack*: The beginning portion of the envelope, it is the time it takes for a sound to reach its peak value (see Figure 6.8). Highly transient, percussive sounds (including pitched percussive sounds like mallets and pianos) have a quick attack; "lyrical" or "legato" sounds generally have a slow attack.
2. *Decay*: The time it takes a sound to "fade down" from the peak value to the value at which it will sustain. The envelope in Figure 6.9 has a long decay; the sound takes longer to diminish before reaching its sustain level. Figure 6.10 is an example of an envelope with a short decay, meaning it reaches its sustain level much more quickly.

Figure 6.8. "Attack" of ADSR envelope.

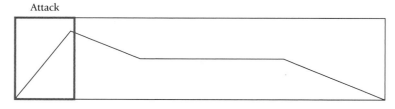

Exploring the Use of Musical Sound Effects

3. *Sustain*: The main value level of the envelope after its initial transient. The envelope in Figure 6.11 "attacks" to a maximum value, then "decays" to a "medium-value" sustain, which lasts until the initiation of the *release trigger*, the time at which the sound is no longer actively produced or allowed (for example, the time at which a clarinetist ceases to blow air and vibrate the reed, or the time at which a finger is lifted off the key of a keyboard instrument). Not all envelopes contain a sustain (that is to say, some envelopes have a sustain length of "zero," thus meaning the decay slope is also the release slope). An envelope without a sustain component, as in Figure 6.12, describes the familiar "transient" sound of most percussion instruments, as well as other musical techniques like *staccatissimo*.
4. *Release*: The final portion of the envelope, it is the time it takes an envelope to reach a value of "zero" after the release trigger (see Figure 6.13). If an envelope has no release (that is to say, if an envelope has a release value of "zero"), then the sound will stop abruptly at the release trigger, as seen in Figure 6.14. However, if it has a long release, it will fade out gradually even after the note is no longer being held down.

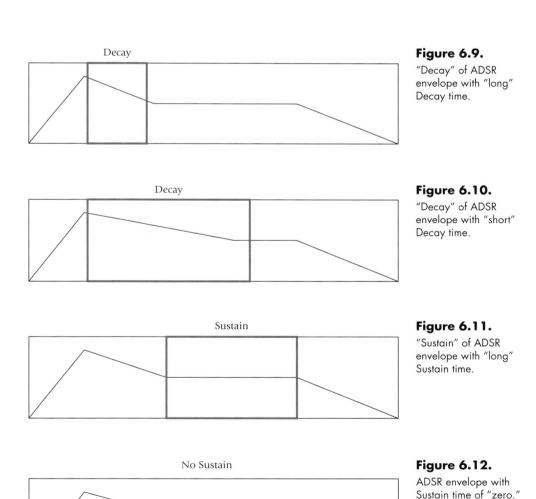

Figure 6.9.
"Decay" of ADSR envelope with "long" Decay time.

Figure 6.10.
"Decay" of ADSR envelope with "short" Decay time.

Figure 6.11.
"Sustain" of ADSR envelope with "long" Sustain time.

Figure 6.12.
ADSR envelope with Sustain time of "zero."

Creating the Music Design

There are two basic kinds of sonic envelopes: *amplitude envelopes*, which describe the relative amount of volume of sound over time, and *timbral envelopes*, also known as *tonal envelopes*, which describe the relative spectral range of sound over time. All sounds may be described as an interaction between the "shapes" of these two envelopes.

Waveforms

Sound is composed of waves. *Musical* (or *pitched*) *sound* is produced by *periodic waveforms*, which oscillate at regular frequencies. The *fundamental frequency* (or, simply, *fundamental*), the lowest frequency of a periodic waveform, determines the pitch of the sound, while *harmonics* (or *overtones*), mathematical multiples of the fundamental frequency, determine how complex the resultant timbre is, as seen in Figure 6.15. Subtractive synthesizers contain one or more *oscillators*, which produce periodic waveforms of one or more of the following types:

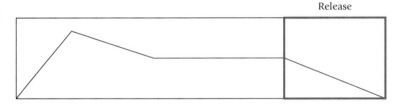

Figure 6.13.
"Release" of ADSR envelope.

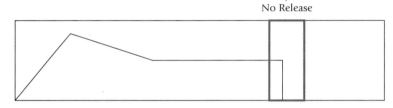

Figure 6.14.
ADSR envelope with Release time of "zero."

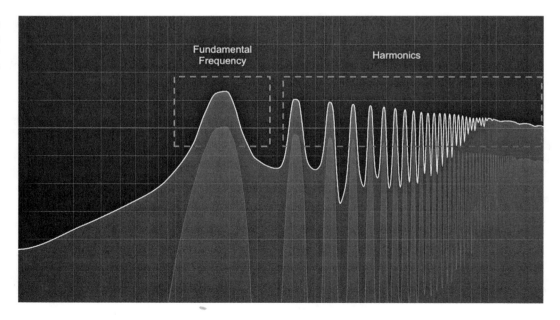

Figure 6.15.
Spectrograph of a fundamental frequency with overtones.

- *Sine wave*: ▶(6.22) The sine wave is the most simple periodic waveform. A true sine wave contains only a fundamental and does not contain harmonics; thus, the sine wave sounds like a simple, pure tone, approaching that of a flute or ocarina (see Figure 6.16).
- *Triangle wave*: ▶(6.23) A triangle wave contains more spectral content than a sine wave, but still retaining a somewhat "rounded" tone. In other words, a triangle wave sounds like a sine wave with some "bite." Because of this, this particular waveform was often used to perform basslines in early consoles, as in the underground theme of *Super Mario Bros.* (1985) for the Nintendo Entertainment System[21] (see Figure 6.17).
- *Square wave*: ▶(6.24) Along with *PWM waves*, the category of waveforms to which it belongs,[22] a square wave has even more harmonic content than a sine wave or a triangle wave. Containing only odd-integer harmonics (that is to say, the fundamental [the first harmonic], the third harmonic [octave + perfect fifth above the fundamental], the fifth harmonic [two octaves + major third above the fundamental], and so on), its sound approaches that of a clarinet in the *chalumeau* register. This particular waveform was often used to perform melodies in early consoles; the iconic melody of the *Super Mario Bros.* Overworld Theme is performed on a square wave from one of the Nintendo Entertainment System's PWM channels, as is the main overworld melody in *The Legend of Zelda* (1987).[23] Likewise on the Sega Master System, the main theme to *Alex Kidd In Miracle World* (1987)[24] exhibits this usage (see Figure 6.18).
- *Saw wave*: ▶(6.25) A complex waveform containing both odd and even harmonics, so named for the "sawtooth" pattern of the waveform (as in Figure 6.19). Bright and brash, it has a tone approaching that of a saxophone or the *cuivré* of cylindrical brass.

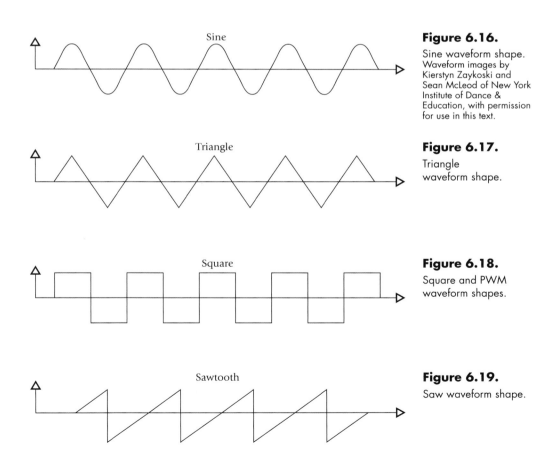

Figure 6.16.
Sine waveform shape. Waveform images by Kierstyn Zaykoski and Sean McLeod of New York Institute of Dance & Education, with permission for use in this text.

Figure 6.17.
Triangle waveform shape.

Figure 6.18.
Square and PWM waveform shapes.

Figure 6.19.
Saw waveform shape.

- *Noise:* ▶(6.26) Itself not a periodic waveform, noise is instead a random signal occurring over a broad frequency spectrum. Most synthesizers have what is referred to as a "noise generator." A noise generator generally produces *white noise*, a random signal having equal power or intensity across all frequencies.[25] Being nonperiodic, noise is necessarily unpitched, so it is often used to create unpitched musical effects like percussion elements or rhythms, as in the drum part to the aforementioned *Super Mario Bros.* overworld theme.

Filters

Filters pass or reject certain frequencies. The four most basic types of filters are named for what parts of the frequency spectrum they pass or reject:

- *High-pass*: A high-pass filter allows higher frequencies to pass through while attenuating lower frequencies below some threshold, as seen in Figure 6.20.
- *Low-pass*: The opposite of a high-pass filter, a low-pass filter allows lower frequencies to pass through while attenuating higher frequencies above some threshold, as seen in Figure 6.21.
- *Band-pass*: A band-pass filter is in actuality a high-pass and low-pass filter acting in tandem, attenuating higher and lower frequencies, allowing a certain frequency band in the middle to pass through unaffected, as seen in Figure 6.22.
- *Band-reject*: The opposite of a band-pass filter, a band-reject filter allows higher and lower frequencies to pass through unaffected but attenuates a certain frequency band in the middle, as seen in Figure 6.23.

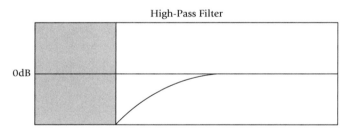

Figure 6.20. Typical high-pass filter EQ effect.

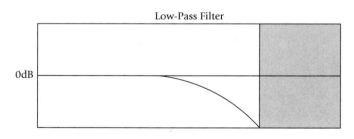

Figure 6.21. Typical low-pass filter EQ effect.

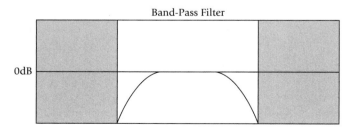

Figure 6.22. Typical band-pass filter EQ effect.

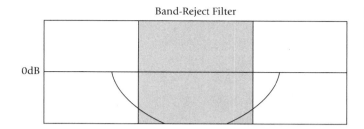

Figure 6.23.
Typical band-reject filter EQ effect.

Other Varieties of Synthesis

Of course, there are a number of other varieties of synthesis available to the modern composer. We will now discuss a few major categories of synthesis that have been used extensively in game music, examining what processes and attributes give each variety of synthesis its own "sound."

Subtractive synthesis, as we have seen, is so named because it involves beginning with a waveform and "subtracting" frequencies, using filters, until the desired sound is achieved. Most classic hardware synthesizers and video game sound chips use this form of synthesis.

Additive synthesis, on the other hand, is the process of adding simple sine waveforms together in various ratios in order to create a more timbrally complex waveform. In other words, additive synthesis involves building timbres by adding overtone frequencies to the fundamental waveform. Many organs work this way.

Frequency modulation synthesis involves one fundamental waveform, the *carrier*, which is modulated by one or more secondary waveforms called *modulators*. A modulator is applied to the carrier, turning it into a new waveform by changing its shape and distorting the amplitude and/or tonal ratios. FM synthesis can be rather economical because, even with only two oscillators, rich waveforms can be produced. However, the effects that modulators can have on carriers are not as immediately intuitive as the modifications subtractive synthesis allows, so understanding this form of synthesis takes more time and effort. Many classic digital synthesizers (such as the Yamaha DX7), the Sega Genesis/Sega Mega Drive, and classic computer sound cards like the AdLib and the Sound Blaster used FM synthesis.[26]

Wavetable synthesis, exemplified by PPG and Waldorf hardware synthesizers and *Massive* (Native Instruments) and *Serum* (Xfer Records) software synthesizers, create periodic waveforms from the repetition and manipulation of an arbitrary single-cycle waveform. Confusingly, the term "wavetable" was also incorrectly used in the marketing and popular discussion of many synthesizers in the 1980s and 1990s; notably, the Korg WaveStation series actually uses wave sequencing and vector synthesis in its creation of sounds.

Summary

Ultimately, whether because of early hardware limitations or modern aesthetic sensibilities, synthesis is a fundamental component of most game scores and soundscapes. The effective composer or sound designer, in understanding synthesis, will be able to create nearly any imaginable sound, shape, or timbre. However, for certain scenarios (such as acoustic realism), another invaluable sound-creation tool is also available: *sampling*.

Introduction to Sampling

While synthesis is the process of designing sound directly from waveforms and tends to produce a distinctly electronic timbral result, *sampling* is a sound-creation process performed by using pre-rendered audio snippets to create playable virtual instruments. As we saw earlier in the chapter, a sample is a short file of an audio recording. A virtual instrument can be anything as simple as one sample pitch-shifted "across the keyboard," to quite complex, using *round robins*,[27] *velocity*,[28] and/or other facets of performance.

In their lecture *Beyond the Presets* at GameSoundCon 2018, Brian White and Brian Trifon of Finishing Move distinguish broad categories of virtual instrument paradigms along this complexity spectrum: A *basic sample kit* is an entire instrument based on the use of one single sample. A *tempo-synced rhythmic instrument* is an instrument with a looping element that adheres to the tempo set in the DAW. A *percussion kit* is built with transient samples with the purpose of being used for percussion. Finally, a *non-chromatic/modal mapped kit* is sampled and programmed to adhere to a specific scale or mode rather than having every chromatic pitch in the range be playable.[29] In addition to those categories, a *sound effects kit* is an instrument that contains various samples of sound effects, and a *complex sample kit* is one that utilizes round robins and/or multiple velocities. *Libraries* are commercially available collections of virtual instruments or kits that can contain anything from a single solo instrument *patch* to a full, deeply sampled orchestra. *Patches* are instruments, articulations, or preset sounds[30] assigned to a particular MIDI channel.

Introduction to Samplers

Sampled instruments are built using *samplers*, hardware or software engines that allow the user to attach samples to specific pitch ranges so that when, say, a key is hit, the relevant sample is triggered. The software sampler Kontakt by Native Instruments is one of the most commonly used samplers in the music industry; let us now observe some organizational and functional elements common to most samplers by observing their implementation in Kontakt.

Figure 6.24 illustrates Kontakt's basic UI. At the top left are five tabs from which different forms of library search and instrument mapping may be carried out. The "Libraries" tab is currently selected, allowing browsing of Native Instruments–licensed sample libraries. The "Database" tab allows granular browsing through the instrument library collection by instrument type, and the "Files" tab allows browsing the computer's file hierarchy for any Kontakt-compatible virtual instrument. On the right side of the window, the *Session Strings Pro 2* library has been loaded.

Figure 6.25 shows some advanced editing features available "under the hood" in the full versions of Kontakt. This view may be accessed by clicking the wrench in the upper left-hand corner of any patch. There are five tabs across the top of this view, each of which opens a sub-window in the UI which allows fine-tuning of nearly any parameter in a given patch. This particular window features the "Mapping Editor," which shows and allows editing of which MIDI notes and velocities trigger which samples—in other words, which samples are "mapped" to which keys.

Exploring the Use of Musical Sound Effects

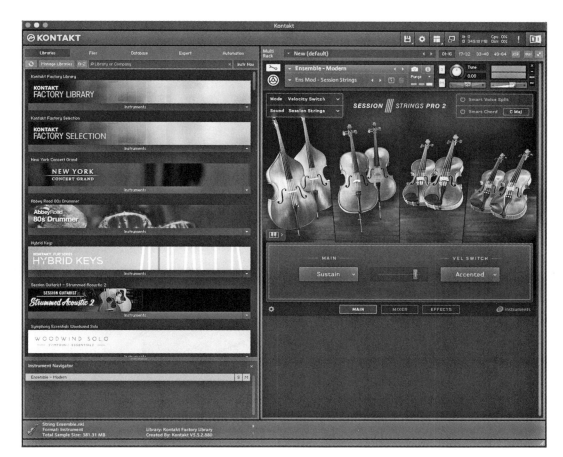

Figure 6.24.
Kontakt 6.2.1 "front page" UI, Kontakt 6, Native Instruments, Berlin, Germany, 2019.

Aside from excellent sample-mapping capabilities, Kontakt also offers tools such as a wave editor, a sample looping editor, round-robin capabilities, numerous audio-processing effects, and even a script editor in which the Kontakt Script Processor programming language may be used for even more possibilities and customization when designing and programming virtual instruments.

While this particular example is of a relatively complex sample kit, creating a basic sample kit is as easy as importing a single sample and defining its range across the virtual keyboard. Based on the root pitch, the Kontakt engine would then automatically transpose the sample by the relevant number of semitones for each key in its range.

Conclusion

Whether by creating synthesized instruments or sampled instruments, designing a set of unique musical sound effects for your soundscape requires thought and time. Effective composers and sound designers would be better served by the deep mastery of one synthesizer and one sampler than a wide but shallow knowledge of many of each; in doing so, entire realms of possibilities for designing original and memorable musical sound effects become easily accessible.

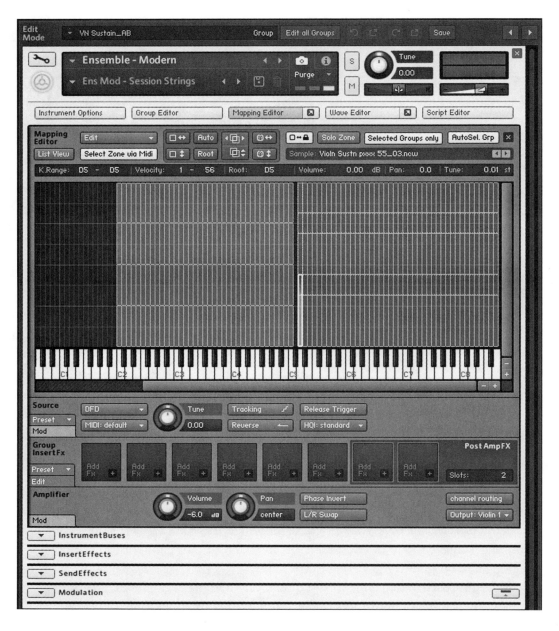

Figure 6.25.
Kontakt 6.2.1 "back page" UI, Kontakt 6, Native Instruments, Berlin, Germany, 2019.

Exercises

1. *Mega Bowling* is a hypothetical bowling game that uses motion controllers to imitate a real-life bowling experience at home. The developers want the player to feel "good" upon successfully knocking down pins. However, they have also pointed out to you that not all shots are equal—there are three levels: "good," "great," and "perfect."
   a. For this exercise, design an instrument from scratch using a synthesizer of your choice.
   b. Then, using that instrument, create a musical sound effect for each of the three shot success levels.
   c. Create a contrasting sound to occur after a complete miss.
2. *Three Blind Mice* is a hypothetical RPG that involves a great deal of dungeon exploration. However, much like *Link's Awakening,* visual cues alone will not suffice

to tell the player where to find an important item. The game contains three types of objects hidden in various rooms: magical weapons, potions, and essential quest items.

   a. Create a new instrument in Kontakt and import one or more audio files, mapping them across the "Mapping Editor" to create a playable sampled instrument.

   b. Use this instrument to create the musical sound effects to indicate to the player where items are located in *Three Blind Mice*, differentiating between magical weapons, potions, and essential quest items.

Notes

1. *Mario Kart 64*, game cartridge, developed by Nintendo EAD (Kyoto: Nintendo, 1996).
2. As we will discuss further in chapter 7, these scenarios in which the music and sound are not directly related to each other aesthetically or sonically, are called *secondary relationships*.
3. *The Legend of Zelda: Link's Awakening*, game cartridge, developed by Nintendo EAD (Kyoto: Nintendo, 1993).
4. Ibid.
5. *Where Shadows Slumber*, mobile, developed by Game Revenant (New York: Game Revenant, 2018).
6. Or, as the saying goes, "killed two birds with one stone."
7. Koji Kondo and Ryo Nagamatsu, "Cavern Theme (Going Underground)," 2013, on *The Legend of Zelda: A Link between Worlds Original Soundtrack*, Nintendo 3679366, 2015, compact disc.
8. *The Legend of Zelda: A Link between Worlds*, game cartridge, developed by Nintendo EAD (Kyoto: Nintendo, 2013).
9. *Donkey Kong Country 2: Diddy's Kong Quest*, game cartridge, developed by Rare (Kyoto: Nintendo, 1995).
10. David Wise, "Bayou Boogie," 1995, on *Donkey Kong Country 2: Diddy's Kong Quest Original Soundtrack*, Nintendo, 1995, compact disc.
11. Released in Japan in 1984 for the Nintendo Family Computer.
12. *Uurnog Uurnlimited*, download, developed by Nifflas Games (Stockholm: Raw Fury, 2017).
13. *The Legend of Zelda: Breath of the Wild*, game cartridge, developed by Nintendo EPD (Kyoto: Nintendo, 2017).
14. *Spyro the Dragon*, game disc, developed by Insomniac Games (Tokyo: Sony Computer Entertainment, 1998).
15. A percussion instrument, also called a *bell tree*, consisting of a number of small, vertically hanging chimes mounted on a bar according to pitch value and performed by sweeping through the chimes in an ascending or descending manner.
16. *The Legend of Zelda: Breath of the Wild*, game cartridge, developed by Nintendo EPD (Kyoto: Nintendo, 2017).
17. *Braid*, game disc, developed by Number None (Redmond, WA: Microsoft Game Studios, 2008).
18. E.g., *Metal Gear Solid*, game disc, developed by Konami Computer Entertainment Japan (Tokyo: Konami, 1998).
19. A type of software that acts as a sound module.
20. In addition to the amplitude envelope, sounds also have a tonal or timbral envelope. Instead of amplitude, the envelope determines the range of tonal frequency.
21. *Super Mario Bros.*, game cartridge, developed by Nintendo Creative Department (Kyoto: Nintendo, 1985).

22. PWM waves are waves with periodic "on" and "off" duty cycles. Waves with equal-length "on" and "off" duty cycles (that is to say, a duty cycle of 50%) are called square waves, whereas waves with uneven duty cycles are called PWM waves. The more uneven the duty cycles are, the more the spectral balance skews toward the upper harmonics, creating a "brighter," "thinner" sound.
23. Released in Japan in 1986 for the Family Computer Disk System.
24. Released in Japan in 1986.
25. Other types of "colored" noise exist, differing based on their relative distributions of energy over the frequency spectrum.
26. All of the examples listed here used Yamaha-branded FM sound chips!
27. A rotating collection of numerous samples per pitch.
28. The speed with which a synthesizer or sampler note is triggered.
29. Brian Trifon and Brian White, "Beyond the Presets," GameSoundCon 2018 (convention lecture), Millennium Biltmore Hotel, Los Angeles, CA, October 9, 2018.
30. Patches can also be found on synthesizers, where the concept first originated; in the earliest synthesizers, before the advent of preset capabilities, particular sounds would have to be built from scratch using patch cables. Later, as synthesizers gained the ability to recall particular configurations of these previously "patched" sounds, the presets themselves became known as *patches*.

7

Music versus Sound Design
Defining the Sonic Relationship

Introduction

As the field of game music continues to grow, so does the importance of sound design as part of the composer's skill set. Each video game has its own visual aesthetic. It is the job of the composer to complement this particular visual world with a similarly unique sonic aesthetic. The relationship between the music and sound design of a game is a central consideration for achieving effective player immersion. Some games seek to achieve immersion by modeling the real world, while others stir the imagination by using sound to enhance a fictional world.

As we saw in Chapter 6, games like *Super Mario Bros.*[1] used musical sound effects to highlight actions or reinforce the player's emotions. For example, the "1 Up" musical sound effect, heard when the player gains an extra life, is a melody in C major (E, G, E, C, D, G) performed on a bright synthesized timbre. The soundtracks of more modern games such as *FEZ*[2] and *The Legend of Zelda: Breath of the Wild*[3] also incorporate musical elements in their sound effects to emphasize emotional content, create atmosphere, and enhance player feedback while solving puzzles.

When the music and sound design of a game are closely related, meaning the sound design contains musical elements that live in the same sonic world as the score, or vice versa, these two core audio elements share a *primary relationship*. Primary relationships can be used for worldbuilding compelling the player's imagination to further construct the world she is in beyond the on-screen visuals. Primary relationships can also be used to inform the player's actions, providing useful sonic clues to aid in overcoming obstacles and moving forward in the game.

However, when a game's developers seek to achieve a greater sense of "realism" in their game, a *secondary relationship* should instead be used. In a secondary relationship, the music and sound design are distinct from one another, similar to the way music and noises[4] are perceived in the real world. Mirroring this real-world behavior, in creating a more realistic aesthetic for the game, fosters a subtler relationship between the music and sound design than that found in a primary relationship.

In the other conceptual direction, when the music and sound design of a game are one and the same, they have a *unified relationship*. This type of relationship is particularly used for creating challenge-based sonic experiences. Games like *Thumper* (2016)[5] use their soundscapes as the primary challenge medium, requiring the player to press buttons and carry out sequences perfectly in-time with the music. *Mini Metro* (2014)[6] uses a unified relationship to enhance the player's experience, creating a sense of progression while also placating the player as the time-dependent challenges grow increasingly more stressful.

Primary, secondary, and unified relationships are thus the three main categories of music-sound design relationships in games. The exploitation and implementation of these relationships play a crucial part in player immersion. Some games require a specific type of sonic relationship implementation in order to be most effective and to create the gaming experience for which the developer is striving.

Thus, in this chapter, since each game has its own set of rules and requirements, we will discuss some different practical examples of these relationship types. We will explore how music and sound effects have interacted throughout the history of game music, and we will define these relationships in basic English. Then, we will look at examples in greater detail and draw some broader applications.

Visualizing Sonic Relationships

Diegesis-Nondiegesis

We will be using the concept of diegesis as a critical determining factor as we discuss different relationship types. Remember, *diegetic* music exists within the world of the game itself, while *nondiegetic* music exists outside of it. However, diegesis-nondiegesis is a continuum, not a binary. Here are two helpful terms that will allow us to distinguish between different diegetic relationships in more detail:

- *Split-diegesis:* The music is completely nondiegetic and the sound design is diegetic.
- *Mixed-diegesis:* The sound design and music overlap in their respective levels of diegesis; in other words, musical elements may exist in the game world itself or may be combined with sounds that exist in the game world.

Now that we have a more granular understanding of diegesis and the vocabulary with which to describe it, let us explore the three levels of music-sound design relationships. Since they are the most common and are arguably the easiest to understand, we will begin with *secondary relationships*.

Music-Sound Design Relationships

Secondary Relationships

Secondary relationships (Figure 7.1) have the following attributes:

- They have a clear distinction between the music and the sound design (split-diegesis).
- They require minimal collaboration between composer(s) and sound designer(s).
- They are often found in games aiming to achieve realism.

In a secondary relationship, the separation between music and sound design is easily discernible. In other words, both categories of game sound exist within the aesthetic of

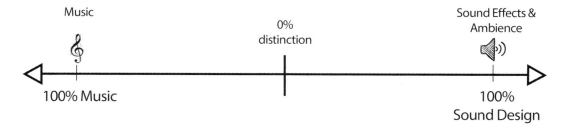

Figure 7.1.
Secondary Sonic Relationship, demonstrating complete separation between music and sound design. Relationship diagrams by Kierstyn Zaykoski and Sean McLeod of New York Institute of Dance & Education, with permission for use in this text.

the game, but are designed with different sonic purposes. As we saw before, this often tends to be the sonic relationship in game types that have environments based on "realism." The sound effects are diegetic, while the music is nondiegetic; this relationship is therefore *split-diegetic*.

Consider how noises are perceived in the real world: walking down the street, one may hear the sounds of cars passing by, their tires rolling across the pavement and their horns blaring, distorting in pitch as they speed past. In a restaurant, one may hear the sounds of people chatting, the clinking and clanking of silverware, the laughter of a loud individual across the room. While these noises might arguably have a sort of musicality, perhaps even an element of tonality (such as the car horns), they are not music; these sounds on the street or in the restaurant do not fit any particular musical scale, follow any specific rhythm, or follow an aesthetically informed temporal organization.

A secondary relationship between music and sound design is distinguished by this type of clear separation of roles. Anybody hearing a recording of a work by Beethoven and a recording of the ambient sounds of a restaurant could easily discern which functions as music and which functions as noise.

For example, the games of the *Call of Duty* franchise are first-person shooters that seek to create a highly realistic atmosphere. The primary goal of the sound effects is thus to foster the player's immersion by enhancing this sense of realism: the sounds of the guns are generally modeled on real gun sounds; most other effects throughout the games likewise reflect the attributes of their real-world counterparts. Even in the *Call of Duty* games with ahistorical storylines, like *Black Ops III*,[7] the audio effects are still tailored to sound as believably realistic as possible for the futuristic reality of the setting. While the sound effects remain completely diegetic, the music is almost always nondiegetic, existing outside of the game and simply supporting the visual image without ever playing an integral role in interacting with the player. The music of the *Call of Duty* franchise is thus more cinematic than ludic in that it complements rather than affects the on-screen actions. ▶(7.1)

Primary Relationships

Primary relationships (Figure 7.2) have the following attributes:

- They have a blurred distinction between the music and the sound design (mixed-diegesis).
- They require extensive communication and collaboration between music composer(s) and sound designer(s) (unless, of course, these are one and the same).
- They are often found in games with fictional game environments.

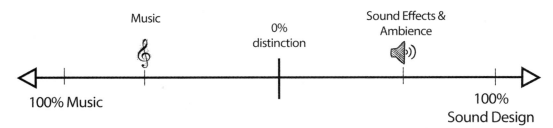

Figure 7.2. Primary Sonic Relationship, demonstrating a closer relationship between music and sound design.

In a primary relationship, the music and sound design, despite their existing as two separate audio entities, often blur together or overlap. The two elements share many aesthetic characteristics and are created using similar, if not the same, tools. In this relationship type, sound effects tend to have musical attributes, and music can sometimes sound ambient or supply part of the designed soundscape; the relationship is therefore *mixed-diegetic.*

For example, Disasterpeace's score to *FEZ* combines classic game sound conventions with contemporary design to create a unique soundscape that embodies the game's dual aesthetic of nostalgia and futurism. Because of the direct role he played in creating the ambiences, it also showcases a composer evolving from writing only music to writing overarching ambiences that blend seamlessly with the sound design of the game, despite there also having been a separate game sound designer (Brandon McCartin). The sound effects are sometimes musical notes that originate from a musical scale. The ambiences function to create not only a sense of reverberant space, but also a tonal musical landscape for the notes to fit into. Designing the ambience himself gave Disasterpeace the freedom to seamlessly blend the ambience and the music and to create musical sound effects that integrated coherently into the soundscape and gameplay.

As the music in a primary relationship may be acting partly as a sound effect attached to, or perhaps emanating from, an object in the game world, it becomes more difficult to ascertain which elements should be defined as music and which should be defined as sound effect. This potential ambiguity is how primary relationships supply unique and creative opportunities for composers to shape the sonic atmosphere of a game. In *FEZ*, for example, the music of the graveyard level mimics the behavior of the thunderstorm climate, employing a musical thunderstorm as the sonic atmosphere. The music system was designed such that Disasterpeace could set a range of time during which each sonic element would play, including starting with a low transient boom that sounded like thunder, and then pausing before the rain enters, much like a real storm. ▶(7.2)

Primary relationships may be further classified into these subcategories:

- *Informative primary*: In this subcategory of primary relationship, musical sound effects are employed only in specific utilitarian cases within the game. For example, many puzzle-platform games (such as *FEZ*) use non-musical sound effects for standard game movements and actions, and reserve the use of musical sound effects for only puzzle-solving actions, thereby giving the player positive feedback that he is on the right track. In the first three-dimensional level of *FEZ,* the player's objective is to pick up a series of yellow cubes that will eventually form one final larger cube once the level is fully explored and every smaller cube has been collected. Almost every sound effect in this level is non-musical except for the sound effects that play when a small cube is picked up—these are musical, even in the same key as the concurrent music score. In a case such as this, musical sound effects are used to inform the player's sense of progress toward the objective of the level. ▶(7.3)

- *Unified primary*: In this subcategory of primary relationship, many or all of the sound effects contain musical elements. This exceptionally intimate connection between the music and sound effects is used most commonly for worldbuilding, strengthening player immersion in the fantasy of the game world by creating a world in which even diegetic sounds are very musical. ▶(7.4) However, even in a unified primary relationship, one can still distinguish between the music and sound effects because their functions are separated. As we will see, the relationship is fully *unified* when there is no distinction between the two: they serve the same purpose. PHÖZ's score to *Where Shadows Slumber* (2018),[8] for example, often uses unified primary relationships for worldbuilding, as shown in the companion site example.
- *Mixed primary*: This subcategory of primary relationship is a mix of the subcategories of informative and unified primary relationships. In this subcategory, there are different types of musical sound effects used throughout the game. Some of them are used for specific reasons or functions, providing the player with important information, while others may be simply used for worldbuilding purposes. Realistically, you are likely to find games that blend different category and subcategory types depending on the necessities of each level. When you are designing the soundscape of a game, you should keep an open mind, reflecting on which category can best serve a soundscape and provide the player with the best possible gameplay experience level-by-level.

Unified Relationships

Unifed relationships (Figure 7.3) have the following attributes:

- They have no distinction between the music and the sound design.
- They require the composer(s) to be the sound designer(s), or to work so closely that the functional distinction between each role is moot.
- They are commonly found in challenge-based games, puzzle games, or games with fictional worlds.

In a unified relationship, there is effectively no distinction whatsoever between the music and the sound design. In fact, in games with such sonic relationships, the composer is almost always also the sound designer.

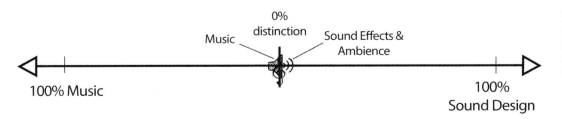

Figure 7.3.
Unified Sonic Relationship, wherein the music and sound design are one.

Mini Metro, a game you may remember from Chapter 5, which features another soundtrack by Disasterpeace, is a clear example of such a relationship. As described by its developer Dino Polo Club on the game's official website:

> *Mini Metro* is a minimalistic subway layout game. Your small city starts with only three unconnected stations. Your task is to draw routes between the stations to connect them with subway lines. Everything but the line layout is handled automatically; trains run along the lines as quickly as they can, and the commuters decide which trains to board and where to make transfers.[9]

The player's objective is to sustain the functionality of the subway system by routing the lines intelligently with the resources available as the game city continues to grow at a steady pace and the game likewise increases in difficulty. The vast majority of the sounds at any given time in *Mini Metro* are based on a single musical chord or scale as predetermined by Disasterpeace. Non-melodic sounds are also generally built upon rhythmic elements aligning with the tempo of the music. When the music and sound share a unified relationship, the player's actions have a profound effect on the resulting audio. In *Mini Metro*, the timing and placement of new subway lines affect when musical tones enter and exit the soundscape, allowing the player to interactively take part in the composition of the music itself. You will have the opportunity to learn more in depth about creating a highlyreactive score of this type when we discuss algorithmic music systems in Chapter 11. ▶(7.5)

Another example of a soundtrack exhibiting a unified relationship is in the aforementioned game *Thumper*. In *Thumper*, the player takes on the role of a beetle trying to escape from Hell. In order to do so, the player must time his or her button presses to be in synchronization with the beat of the music. The sound effects that play when a particular beat is hit also land in time with the music and have an instrumental sonic aesthetic quality. The entire soundtrack is therefore tied directly to the music and its rhythmic timekeeping. ▶(7.6)

Unified relationships are fairly uncommon in AAA games. However, as the Indie game market continues to expand, so does the opportunity for composers and sound designers to create adventurous audio experiences that blur, or even completely erase the lines between music and sound design.

Summary

As shown in Figures 7.1, 7.2 and 7.3, the relationship between music and sound design can be imagined as belonging on a spectrum; the particular degrees to which each element falls along that spectrum is unique to each particular game. Games with secondary relationships place music and sound design at opposite ends of the spectrum with complete distinction between the two. This relationship type is useful for hyper-realistic games since the sounds we hear in the real world do not, for the most part, contain any musical elements or devices. A primary relationship, on the other hand, blurs the lines between which sounds fit which category, giving the composer the opportunity to worldbuild with musical elements attached directly to game objects or gameplay. In a game with a unified relationship, the music and sound design both fall directly in the center of the spectrum, with 0% distinction. In reality, most games will have differing levels of distinction, and this might even change depending on the game level or scene and how the composer or sound designer has chosen to craft the sonic relationship for each specific situation.

Creating a Timbral Landscape by Defining the Relationship

Designing a Primary Relationship: A Closer Look at *FEZ*

A game like *FEZ* poses the challenge of combining multiple, seemingly conflicting ideas into one cohesive aesthetic unit. In this case, a primary relationship is employed in the soundtrack to help achieve this goal, successfully using audio to worldbuild. Both the music and sound design sonically mirror the "modernized low-bit" graphics and convey the combined senses of futurism and nostalgia inherent in both the game's overall aesthetic and within the lore of the story.

The futurist aspect of the game comes from the folklore of an ancient alien civilization living in a multispecies universe. As described by Studio Polytron's official *FEZ* website, in *FEZ* the player plays "as Gomez, a 2D creature living in what he believes is a 2D world [. . . until] a strange and powerful artifact reveals to him the existence of a mysterious third dimension!"[10] The most integral game mechanic in *FEZ*, switching perspective from 2D to 3D, literally crosses the dimensional barrier. However, an element of nostalgia is also presented as the graphics heavily reference the retro, low-bit visual aesthetic (such that there is even a facsimile of a poster of the iconic title screen of *The Legend of Zelda* [1986],[11] as seen in Figure 7.4). Since the futuristic alien civilization is actually ancient, this creates a sense that everything in the game is very old; the ghost of a memory.

In a 2013 lecture at *Gamer's Rhapsody*, Disasterpeace notes that it is essential for every composer to have at least one single mastered tool in his or her arsenal.[12] Once this tool is mastered by the composer, it can function in the creation of sounds, ambiences, and musical textures. In a case like *FEZ,* where music and sound design would share such a close relationship, such a versatile tool was essential to the creative workflow. Disasterpeace's tool of choice was Native Instruments' digital synthesizer *Massive,* which he knew inside and out and was able to use to achieve the majority of the sounds he wanted.

Figure 7.4.
The Legend of Zelda cover as a poster on the wall in *Fez*.

Most DAWs have at least one built-in synthesizer that can be used for this very purpose. *Logic Pro X*, for example, contains a variety of powerful synthesizers that come with the program, like *ESM Monosynth, ESP Polysynth, Retro Synth, Sculpture,* and *Alchemy*; as does *Ableton Live*, with *Analog* and *Operator*. There are also many powerful third-party synthesizer plugins, too, like *Razor* by Native Instruments or the *V-Collection* by Arturia. By mastering the use of even one synthesizer, composers allow themselves the opportunity to create any number of sounds with a single tool. This is a powerful and effective creative position to be in when designing a primary sonic relationship.

The musical score to *FEZ*, composed largely using *Massive*,[13] is proof of how much the mastery of a composition tool can add to a composer's arsenal. Disasterpeace created many unique effected sounds, all within *Massive*. For example, he often placed *Massive*'s reverb on his original patches to change the evolution of the sounds.[14]

Since the majority of the *FEZ* soundscape is rooted in the sound of low-bit synthesis, the tools specific to *Massive* gave Disasterpeace the freedom to design both the music and ambience such that it shared its sonic aesthetic with the sound design. Almost all of the sounds originate from simple waveforms and noise generators. Noise layers create background motion and a sense of atmosphere, sometimes in the music and sometimes in sound effects. The "ancient" speech sound effect, spoken by alien characters throughout the game, is layered with a low-bit square wave, giving the speech an extra element of sonic texture as if more than one set of vocal cords were vibrating. Throughout the game, filter sweeps and LFO modulation in the music and ambience bring sounds to life as they progress and grow; for example, in the track "Rise," a flanger effect is applied to certain pieces of the music, contributing an element of futurism due to its clearly digital sonic nature. At other times, the desampling (decreasing in sample rate and/or bit depth) of a sound portrays the game's retro, low-bit aesthetic. A prime example of this phenomenon is when the 2D world in which players begin their journeys visually disintegrates into many little images.

During his *Gamer's Rhapsody* talk, Disasterpeace also refers to two of what he calls "secret sauces." The first is his use of *Logic*'s *Bitcrusher* bit-crushing distortion effect plugin, which he used on almost every cue in *FEZ* in order to give the music a more authentically retro low-bit sound, which also helped to shape the music's aesthetic similarity to the sound design. He generally placed the bit crusher on the master bus, subtly de-sampling the sound of all of the cues, squashing their fidelity and revealing the sound of the noise floor, making them evoke the sound palette of an actual retro system.

Disasterpeace's second "secret sauce" is the use of a pitch LFO. Because older waveform oscillators were analog, they tended to de-pitch naturally over time, creating a subtle but important variance in pitch that became part of the "warmth" of analog synthesizers. This effect, though now largely controlled or emulated by digital means, continues to produce the same warmth in modern synthesizers. In order to reproduce this effect, Disasterpeace used the *Logic* plugin *Tape Delay* as a pitch modulator, manipulating its "wow" effect. Even though the oscillators used in early gaming hardware sound chips did not have this pitch-drifting problem to the extent that many classic analog hardware synthesizers did, Disasterpeace's use of this subtle sonic modulation still evoked the sound of a particular era of pop culture awareness.

In a fantastical game with a visual aesthetic revolving around pixelated digital landscapes, the music and audio must logically complement the visual aesthetic in order to enhance player immersion. In the case of *FEZ*, Disasterpeace was also in charge of creating the *sonic ambience*. Generally speaking, music and ambience are often rather different sonic elements, fulfilling different roles and functions. While music might hint at the size or the characteristics of a location in a film, rarely does it create the diegetic ambience of that location. In *FEZ*, however, the music plays such a central role in the sonic environment that the essence of the location is "built into" the music itself. On this level of integration, it becomes difficult to even call such a product "music"; the argument that it could be categorized as "ambience" is just as strong. The difficulty of definitively defining and categorizing this soundscape is an effect of the primary relationship inherent in the *FEZ* soundtrack: a marriage of music and sound design. The aesthetic similarities between the music and sound design, strengthened through the use of *Massive* to create pieces of both elements, create an effective complement to the compelling and unique aesthetics of the game's visuals.

Crafting a Unified Relationship

The greatest advantage of the use and implementation of a unified relationship in the soundtrack of a game is that it allows for a uniquely interactive experience for the player. Because each sound effect is tied directly to the music, most player actions will have a direct effect on the music. Therefore, the player could notice and act on patterns of how his or her gameplay actions could influence the music. In other words, the player is allowed to assume a central role in the creation of the soundscape because of how directly their actions affect it.

However, ceding this creative power to the player in a poorly designed game could be detrimental to the gameplay experience. For example, imagine that a player discovers that, by carrying out a specific series of actions, he can create a pleasing musical pattern of sound effects. In fact, one of the most enjoyable elements of the gameplay for him is triggering this pattern. If the game lacks thoughtful design, this behavior can actually lead to the player achieving a lower score or even losing the game—in other words, a suboptimal or even harmful gameplay action has been accidentally incentivized.

On the other hand, if the game is designed well, the sound will affect the player in one of two ways: it will either encourage actions that lead to winning behavior, or it will remain completely uninfluential toward the outcome. Each sound is purposeful and intentional. Therefore, when designing a unified soundscape, a composer must be acutely aware of how compositional and implementation decisions will affect the player.

Disasterpeace's aforementioned *Mini Metro* score is successful at creating a sonic atmosphere that relaxes players, but does not directly influence their decision-making. Altering subway lines often leads to changes in the music; however, these changes rarely have a tangible effect on the overarching musical harmony. Rather, it tends to be the sounds which are out of the player's control that create the larger changes; for example, bass notes enter and exit, changing the tonic of the current chord. However, the player has no direct control over what these root notes are.

In a 2016 interview for *Designing Sound*, Disasterpeace describes the audio functionality in *Mini Metro* in the following way:

Each city/level has an inherent set of musical qualities; access to certain rhythms, harmonic choices, and train engine sounds among other things. Which of those are played at any given time is somewhat controlled by the player through their decisions around the size and shape of their subway system. Each level contains strings of notes that represent the harmonic structure and the voice leading of the music. Each subway line has a rhythm and a note tied to it at any given time. Altering the subway line least recently altered replaces the oldest note in the harmonic structure with the next one in line. The rhythms work in a similar way. Altering a line alters the rhythm by shifting it to the next available rhythm in a list of possible rhythms. Sometimes the harmonic structure of the music will change depending on which week of gameplay you are currently in.[15]

As each subway line has its own associated rhythmic value and note value, the player's actions consistently and predictably affect the music. Usually, the player input causes changes to note value and not harmonies. Because of this, the player's influence on the music, while still direct and auditorily cohesive, still gives the player the mental space to concentrate on the primary task at hand: designing a logical subway system.

On a technological level, crafting a unified relationship requires a certain toolset that may not be necessary for other types of sonic relationships. Given that the music and sound are one and the same, each sound should happen in time with the tempo. In the case of *Mini Metro*, the game was developed in *Unity*, one of the most commonly used game engines today. We will discuss game engines, with a focus on *Unity*'s functionality, further in Chapter 13. Within the *Unity* engine, a plugin called *G-Audio* was used to execute sounds at specific rhythmic intervals, allowing the developer to time the playback of sounds so that it would align musically.

Ultimately, the art of crafting a unified relationship for a game requires thoughtful planning and sensitivity to how the player could and will interact with the sound. On a technical level, it also requires smart system design to ensure that the soundscape functions properly throughout the game. As composers, we are best positioned to accomplish the task of creating such a system when we have at least one musical sound creation of choice with which we are highly proficient.

Conclusion

Different categories of sonic relationships offer various advantages and disadvantages to the player, the composer, the sound designer, and the developer. Since each category encourages particular styles of gameplay, and thus might not suit every style of game, proper forethought and planning must be done to find the most effective and appropriate sonic relationship for each game. However, though a "realism"-centric game would probably fare better with a secondary relationship than with a primary relationship, and a fantastical, abstract game would probably fare better with a primary or unified relationship than a secondary relationship, ultimately, the composer, the sound designer, and the developer should think carefully about each and every game they develop, taking advantage of the wonderful opportunities the right relationship can afford.

Exercises

1. Choose a game that displays either a primary relationship or a unified relationship. Define the relationship: how do the music and sound effects in the game interact?
2. After doing some research, determine how the composer and sound designer created the soundtrack. What tools did they use? Did they use the same tools for creating both elements? Did they collaborate often, or work mostly independently from each other?
3. How is diegesis treated in the game?
4. Why do you think this type of music and sound relationship was chosen for this particular game? Describe the game mechanics and how the player interacts with the game. What are the advantages of this type of relationship? Are there ways in which it detracts from or is limiting to the gaming experience?
5. If you were to score this game from scratch, what would you do differently?

Notes

1. *Super Mario Bros.*, game cartridge, developed by Nintendo Creative Department (Kyoto: Nintendo, 1985).
2. *FEZ*, game disc, developed by Polytron (Montreal: Trapdoor, 2012).
3. *The Legend of Zelda: Breath of the Wild*, game cartridge, developed by Nintendo EPD (Kyoto: Nintendo Company, 2017).
4. A non-musical sound. Not to be confused or conflated with the spectral phenomenon discussed in Chapter 6.
5. *Thumper*, game disc, developed by Drool (Providence, RI: Drool, 2016).
6. *Mini Metro*, download, developed by Dinosaur Polo Club (Wellington, New Zealand: Dinosaur Polo Club, 2015).
7. *Black Ops III*, game disc, developed by Treyarch et al. (Santa Monica, CA: Activision, 2015). The game takes place in a hypothetical 2065, wherein the player plays as a cybernetically enhanced soldier fighting in a dystopian, post–climate change world.
8. *Where Shadows Slumber*, mobile, developed by Game Revenant (New York: Game Revenant, 2018).
9. Dino Polo Club, "Mini Metro," published 2016, accessed October 9, 2019, https://dinopoloclub.com/games/mini-metro/
10. Polytron, "FEZ," published 2012, accessed October 10, 2019, http://fezgame.com
11. *The Legend of Zelda*, game cartridge, developed by Nintendo EAD (Kyoto: Nintendo, 1986).
12. Rich Vreeland, "Music Workshop—FEZ," YouTube video, 44:22, November 20, 2014, https://www.youtube.com/watch?v=PH04VJ8jxvo.
13. Native Instruments' virtual synthesizer *Massive* is an excellent example of a well-rounded synthesis tool that can be used for both music composition and sound design. It features three separate waveform oscillators, a noise generator, two filter modules, and two LFOs, as well as multiple effects that can be applied. Some examples of these effects are sample-hold, bit-crusher distortion, delay, and reverb. A composer could easily use *Massive* to design a sound that is both "mechanical" and "musical" in nature, and could even apply ambience to this sound afterward—all within the plugin itself!
14. Rich Vreeland, "Music Workshop—FEZ," 2013.
15. "The Programmed Music of 'Mini Metro'—Interview with Rich Vreeland (Disasterpeace)," Richard Gould, *Designing Sound*, published February 18, 2016, accessed May 7, 2019, designingsound.org/2016/02/the-programmed-music-of-mini-metro-interview-with-rich-vreeland-disasterpeace/.

8

Designing Interaction between the Player and the Music

Introduction

Understanding the different types of relationships between music and sound design is not about merely following a step-by-step guide, but rather using your knowledge of the possibilities and functionalities to creatively implement a creative audio system. As we saw in Chapter 7, crafting and working within the proper sonic relationship is an essential part of any music design. Poor planning can lead to distracting audio, which can inhibit immersion; an appropriate implementation can bring an alchemical *je ne sais quoi*, greatly strengthening immersion.

In addition to the music-sound design relationship, the player will often have an intricate and interactive relationship with the game audio, and understanding and finessing the different ways in which this relationship might function is a crucial part of the music design process.

In this chapter, we will focus our examination on that very relationship, refining our understanding of music and player interaction. We will then explore five categories, or *schemes*, of interaction: *nondiegetic state-driven interaction*, *nondiegetic player-driven interaction*, *diegetic state-driven interaction*, *diegetic player-driven interaction*, and *multi-driver interaction*. Knowledge of these player-music interaction schemes will inform effective game audio system design.

Player-Music Interaction

There are a number of ways in which the player and the music might interact in any given game scenario. First, the player may have either an *indirect* relationship with the music (that is, the player's actions *do not* directly lead to clear musical consequences) or a *direct* relationship with the music (that is, the player's actions and decisions *do* result in clear musical consequences). These interactions might also happen *diegetically*, or *nondiegetically*.

It is important to determine early in the audio-planning process whether or not the player will share a direct relationship with the music. In other words, should the music be fully *interactive*, or should it be merely *reactive*? When an interaction is *state-driven*, the player cannot interact directly with the music; rather, the player interacts with the game state, which in turn causes the music to react. In cases such as these, the music is *reactive*.[1]

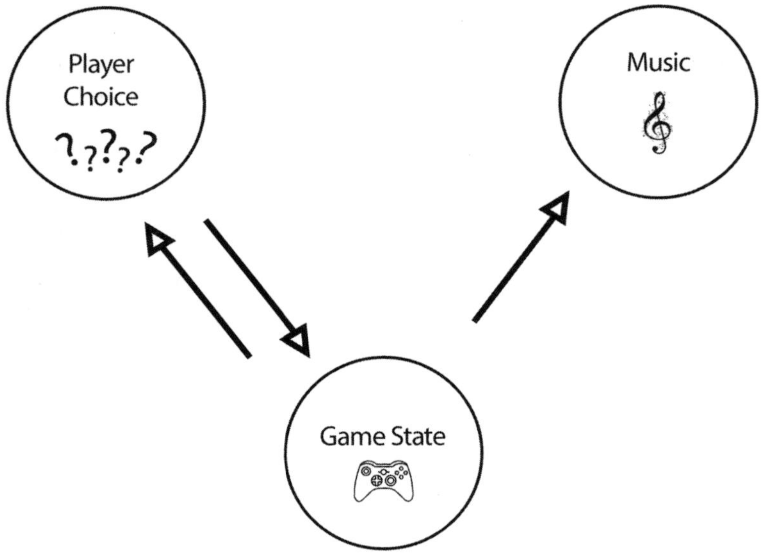

Figure 8.1.
State-Driven Interaction: Player choice changes the game state, which then changes the music. Player interaction with the music is indirect. Game State diagrams by Kierstyn Zaykoski and Sean McLeod of New York Institute of Dance & Education, with permission for use in this text.

In the state-driven relationship scenario outlined in Figure 8.1, let's say that the player has made a decision which has caused the game to transition from an "exploration" state into a "battle" state, which then initiated "battle music." The player's actions have thus affected the game state, and the game state in turn has affected the music. The player's choices did ultimately affect the music, but it was in an indirect manner.

When the music interaction is *player-driven,* the player can affect the music directly without any change in game state occurring; music that shares a direct cause-and-effect relationship with the player in this manner is *interactive*.[2] Unlike a state-driven score, which will change only when the game state changes, an interactive score will react immediately to player actions (Figure 8.2). In such cases, though the player has not (for example) moved closer to an enemy or changed locations, the music has been affected nonetheless. If, for instance, the programmer writes "when the 'x' button is pressed, change musical key," the player can easily press the "x" button at an interval of the player's choice, changing the musical key over and over, even if the game state never changes from "combat" the entire time.

As enticing as the concept of direct player-music interaction may be to consider when planning game audio implementation, there are numerous cases in which a player-driven relationship might actually detract from the immersion. For example, most highly realistic games would not benefit from such a relationship. Take our immersion example from Chapter 4, *Call of Duty: WWII* (2017),[3] which has as its fundamental function the simulation of World War II battles; most real soldiers would not be hearing music during such battles. If the underscore of *Call of Duty: WWII* were player-driven, the music would therefore be a distraction to the players, adding an unnecessary game mechanic that would take their minds away from the war activities and instead encourage them to interact with the music. In other words, the interactivity would encourage the wrong type of gameplay, detracting from the intended gaming experience. This scenario

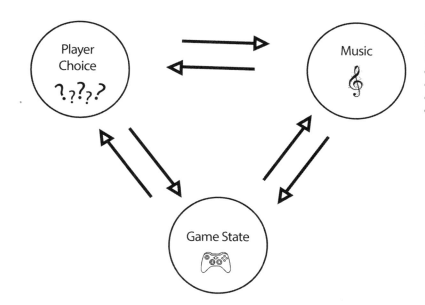

Figure 8.2.
Player-Driven Interaction: Player choice, game state, and music all have a direct, reciprocal effect on each other.

clearly demonstrates why composers must be careful in how they implement the music system to best create immersion.

The Three Dimensions

In the previous chapter, we explored the different relationships that music can share with sound design. Now, we will dig more deeply into these sonic relationships while addressing yet another crucial element of our conceptual toolset: how the player will interact with the sound.

Each player-music interaction scheme may be defined by three fundamental characteristics:

1. *Diegesis*: Is the music *diegetic, nondiegetic*, or some combination of the two?
2. *State*: Does the player have *direct* or *indirect* control over the music?
3. *Relationship*: Is the sonic relationship *secondary, primary*, or *unified*?

Let us now explore some of these resultant combinations.

Nondiegetic State-Driven Interaction

In games with secondary sonic relationships, state-driven interactions are incredibly common. The music is nondiegetic and remains audibly separate from the sound design in the soundscape, strengthening immersion instead by subtle variations in reaction to transitions between game states. The player has no direct interaction with the music. Rather, the music changes when an event changes the state of the game. Thus, this scheme is a *nondiegetic state-driven interaction*.

Horizon Zero Dawn (2017)[4] is an example of a game that successfully employs a nondiegetic state-driven interaction scheme. During gameplay, the player moves between

Creating the Music Design

exploration, stealth, and combat. As one approaches a group of enemies, the game state may transition from exploration to stealth, the music intensifying accordingly. The score (which, aside from a few minor instances, remains completely nondiegetic throughout the game) changes state yet again if the player enters battle. However, the player does not (and cannot) interact with the music directly.

At times when the player (as Alloy, the protagonist) is in a stealth state prior to an enemy being alerted to her presence, the music generally remains silent, allowing the sound design of the landscape and the machines trouncing around to occupy the soundscape. Figure 8.3 illustrates an example of such a scenario.

When enemies hear a noise or notice Alloy nearby, they enter an alert state, the circles above their heads turning yellow with a question mark to signify that they are still unsure if they have detected a threat. The music reflects this, comprising a thin, high-frequency layer of electronic percussion that builds tension and anticipation, while saving the low frequencies for more intense percussion to take over when combat begins. In Figure 8.4, Alloy has just been discovered by the enemy machine, triggering a transition to an enemy alert state in which the enemy is now searching for a threat.

When enemies are in a combat state, the circles above their heads turn to red and show an exclamation mark. In Figure 8.5, Alloy has fully entered battle with this machine, and the music completes its transition into a synth-heavy, up-tempo combat track featuring intense, low drums. Our companion site example illustrates this full transition. ▶(8.1–8.3)

Figure 8.3.
Alloy in stealth state. Screen capture from *Horizon Zero Dawn*, game disc, developed by Guerilla Games (San Mateo, CA: Sony Interactive Entertainment, 2017).

Designing Interaction

Figure 8.4.
Alloy at enemy alert. Screen capture from *Horizon Zero Dawn*.

Figure 8.5.
Alloy in battle. Screen capture from *Horizon Zero Dawn*.

Nondiegetic Player-Driven Interaction

In games in which the player has direct influence over the nondiegetic music, there is a *nondiegetic player-driven interaction*. This scheme is fairly uncommon because of how it clearly neglects the lines of diegesis. In general, nondiegetic music exists as underscore and is state-driven; having the player directly interact with it brings the underscore to the forefront and gives the player an unusual level of control for the state and diegesis type. It is for this reason that nondiegetic player-driven interaction is most often found in "music" games, like the renowned *Guitar Hero* series.[5] In music games, the music is the primary focus of the gameplay, requiring the player to interact with it directly to complete musical tasks, such as playing songs or executing rhythms.

In many cases, the music itself becomes a determining factor in the non-musical game results. *Rez* (2001)[6] is a rail shooter[7] with a core game mechanic of shooting enemies on screen. Each action in the game is attached to the rhythm of the soundtrack. Thus, each shot is in rhythm and results in sound effects that contribute directly to the music melodically, harmonically, and rhythmically. ▶(8.4) The music occasionally even evolves to the beat as a direct result of the player's actions. The soundscape in *Rez* is unified; the music and sound effects are all part of a single whole.

There are also music-based games that use generative music (music generated algorithmically in real time) techniques for the creation of their soundtracks. In such a game, the player (by her actions) has such granular control over the soundtrack that she is, in essence, extemporaneously creating the soundtrack. We will discuss generative music more in Chapter 10, but for now it is sufficient to understand that in this game type, the developer and composer build the sound engine such that the game itself functions almost like a musical instrument. Take, for example, the 2005 game *Electroplankton*,[8] which is centered around 10 different marine creatures and their environments. The player controls both the creatures and their environments in order to create a musical soundscape, creating the score in real time within the confines of the mechanics and systems predetermined by the developers. This game (and games of this type) blurs the lines of diegesis. The musical results are clearly tied to the environment, yet they also feel as though they are, in fact, nondiegetic score. This is, in part, because there are also less-prominent, diegetic environmental sounds as well, which seem to exist separately from the music, further strengthening the impression that the music is, indeed, nondiegetic.

In Figure 8.6, a nondiegetic rhythmic beat sounds in the music score. The player's actions directly influence the tempo of the beat to speed up and slow down in real time, depending on which color of fish is being tapped. ▶(8.5)

Nondiegetic player-driven interaction may be found in more than just music-based games. Imagine, for instance, a game in which the player picks up a magic hammer that changes the background music's key center each time it is swung. The player can swing this hammer at the press of a button, thereby interacting directly with the music, since the music changes its key directly with each press. The music is nondiegetic, sounding only in the background; however, players do have an interactive relationship with the music;

they can affect it directly through their actions. Unlike *Electroplankton*, which gives the player some control over a generative composition, the music in this example would instead feel like a personalized pre-composed soundtrack for the player—in other words, as opposed to a typical game's underscore, which follows the ebbs and flows of predetermined gameplay mechanics and state changes, this form of nondiegetic player-driven interaction would lead to a score that sounds pre-composed, yet also exceptionally reactive.

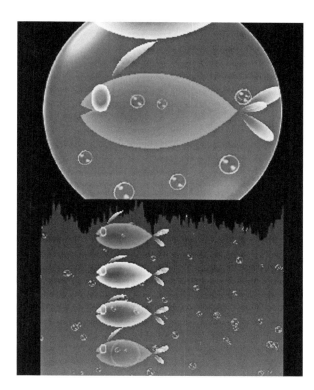

Figure 8.6.
Fish of different colors show where to tap next. *Electroplankton*, game cartridge, developed by indieszero (Kyoto: Nintendo, 2005).

Nondiegetic player-driven interaction can also lead to fresh gameplay results when the sonic relationship of a game is primary. For example, imagine a scenario in which the player must get past a deadly machine which swings an axe directly in tempo with the music. In this case, the music is not determined by state changes, but rather, each time the player presses the jump button, both the music and the machine's axe-swinging rhythm increase by one beat per minute. In such a case, the player would be interacting directly with the music, having control of the music's tempo. The increasing speed of the music would also signal an increase in the potential danger level. Thus the music could be used to serve multiple purposes: to signify game information to players and to interact with them.

Portal 2 (2011)[9] exhibits very creative uses of nondiegetic player-driven interaction. Aside from the diegetic sound effects, the game features a reactive music score that responds directly to many of the player's actions. For example, the game features a mechanic called "repulsion gel," a blue paint-like substance that bounces the player-controlled robot upon contact, repelling it. With each bounce, the score sounds a chordal arpeggiation that changes the underlying harmony. ▶(8.6) In another such creative use, lasers each produce their own musical tone as part of the score. Each time the player changes the direction of a laser or creates a new one, a new synthesizer texture emerges, playing a different musical tone. ▶(8.7) The overall result of this highly reactive system is a game in which the score becomes personalized to each gameplay.

Ultimately, nondiegetic player-driven interaction, especially when the music and sound design are so closely related, clearly allows a certain amount of creative freedom. It can be used as anything from a primary gameplay mechanic, as it is in many music-based games, to a specialized immersionary aid, as it is in some more "realistic" games.

Diegetic State-Driven Interaction

In games in which the player has indirect influence over the diegetic music, there is a *diegetic state-driven interaction* scheme. Diegetic state-driven interaction is more rare than other types typically found only under very specific circumstances within a game—most commonly, perhaps, when characters in a non-music game perform music. Since the music is being performed by an NPC (non-player character), the player character usually has no control over the musical outcome, but the player's decisions can cause state changes that in turn can cause the music to start, stop, or otherwise change.

In *The Legend of Zelda: Breath of the Wild* (2017),[10] the player (as protagonist Link) can travel to Rito Village where he or she will find Kass, a birdlike humanoid creature who performs songs on an accordion throughout the game. Kass wishes to play Link his "Teacher's Song," feeling it is important for him to hear. Whether or not to listen to this song is a choice that the player can make; agreeing to hear it changes the game state and causes Kass to play the song (Figure 8.7). After this, encountering Kass at night affords the player an opportunity to trigger a similar series of events, resulting in Link hearing the music again (Figure 8.8). The player cannot interact directly with the music in this scenario, but did initiate a state change, which in turn caused the music to play in the game world—an example of diegetic state-driven interaction.

Figure 8.7.
Link meets Kass, who plays the "Teacher's Song." *The Legend of Zelda: Breath of the Wild*, game cartridge, developed by Nintendo EPD (Kyoto: Nintendo, 2017).

Figure 8.8.
Link must find Kass at night to hear the "Teacher's Song" again.

Diegetic Player-Driven Interaction

In a game featuring a *diegetic player-driven interaction* scheme, the player can directly interact with a musical element that exists within the game world. This scheme is fairly common for in-game musical puzzles wherein the player must interact directly with a sound or an instrument in the game world in order to advance the character's storyline or to complete a task. In *Bendy and the Ink Machine: Chapter Two: The Old Song* (2017),[11] for example, the player must solve a puzzle by playing the correct sequence of notes in order to open up a new area. To solve the puzzle, the player must play—in the correct order—a dusty old piano, a violin sprawled across a chair, and a rackety acoustic bass. ▶(8.8)

Multi-Driver Interaction

In games featuring a *multi-driver interaction* scheme, there are in fact multiple concurrent interaction schemes: some elements of the music react only to the game state; other elements interact directly with the player. In other words, the music comprises both *player-driven* and *state-driven* elements.

Multi-driver interactions are quite common in games that use music as a core game mechanic, particularly those with unified sonic relationships. For example, in our unified relationship exemplar *Thumper*,[12] the player's actions directly affect the nondiegetic music.

The player's successful navigation of musical rhythms leads to diegetic sound effects which fit perfectly in time with the percussive nondiegetic music. However, the nondiegetic music is reacting only to changes in the game state, and not directly to the player's actions.

Returning to our earlier hypothetical "swing the magic hammer" game, now imagine this: in addition to the player's hammer switching the key of the music with each press of the "hammer" button, entering a battle *also* directly changes the music—this time, to a completely different musical cue. In this scenario, there is one element that is controlled directly by the player (i.e., the musical key), and one element to which the player's actions are only indirectly related via a game state change (i.e., the cue): a multi-driver interaction scheme.

Uurnog Uurnlimited (2017),[13] a game that we will examine in more detail in Chapter 11 during our discussion about advanced algorithmic music systems, is a relatively recent example of a game that successfully employs a multi-driver interaction scheme. Though the music and sound design are deeply correlated, there is also a nondiegetic layer of music that is playing throughout the game; the player can carry out small actions that directly influence and change this non-diegetic music. For example, if the player smashes a box, the music's melody might completely invert, or perhaps the cue might change key. However, some of the sound effects in the game are also musical sound effects, triggered by small or simple player actions. Thus, the player has direct interaction with both the diegetic and nondiegetic layers of the music.

This interaction scheme can be incredibly fun or rewarding for the player, but it is not necessarily the most effective scheme for all types of games. In the case of *Uurnog*, the developer, Nifflas, chose to make the music highly interactive as a core feature of the game itself.[14] In this case, the high degree of reactivity only strengthens the gameplay. However, using such an intertwined relationship between the player and the music can also detract from any game in which enthralling musical interaction would be a distraction from the story or objective at hand. Games designed to achieve realism are a good example of those in which giving the player the power to control nondiegetic musical elements directly could actually draw the player's attention away from the core gameplay mechanic, encouraging him to play with the music instead.

Summary

All in all, a player's interaction with music often differs not only from game to game, but even from moment to moment within a single game. As we have previously discussed, *Horizon Zero Dawn* is a game that uses a *state-driven* secondary relationship for the majority of its soundscape. *The Legend of Zelda: Breath of the Wild* is a game that uses various music-to-sound relationships and different schemes, depending on what is needed in specific scenarios. In games with massive scopes like these, composers have a great deal of opportunity for experimentation with different tools and techniques. Now that you understand all of the different ways in which the player interacts with the music of a game, be sure to take these into account when creating the music design for your project.

One fantastic example of a game that uses a variety of creative interaction techniques is *Inside* (2016).[15] The score often blurs the lines of diegesis, applying different player-to-music relationship types in combination with thoughtful audio implementations to heighten the player experience. Let's explore this more in depth.

Inside: Using Player-Music Interaction Schemes to Enhance Immersion

Inside is a 2D puzzle-platformer game that utilizes multiple different player-music interaction schemes depending on the experiential goals it needs to achieve in each specific scene. Aside from augmenting the game's visually stunning and immersive environment, music is often used to provide important signals to the player. The puzzles in *Inside* require not only mental acuity, but also excellent hand-eye coordination. This type of gameplay creates a serious potential issue for player interaction: the player may successfully discover, and even attempt, the correct solution to a puzzle, and yet still fail because a button was not pressed at the perfect moment. In such a case, it may be easy for the player to assume that her intended solution to that puzzle was incorrect, and therefore continue to waste time while trying to find a new solution, when all that would have been required was better timing of a button press.

To solve this problem, a creative musical approach was used. Rather than employing nondiegetic music in the background at all times, the designers chose to reserve it only for important moments during which a specific message would be conveyed to the player. Most of the time, diegetic musical elements are added into the soundscape, creating a primary sonic relationship. Using nondiegetic music sparingly fosters a situation in which the entrance of music becomes immediately noticeable; this music is a critical clue that may save a great deal of the player's time, if not his in-game life.

Thus, in *Inside*, music is used to signal that a puzzle's solution has been correctly deduced by the player, even if the player has not yet successfully executed the actions that would lead to 100 percent completion of the puzzle. In other words, each time the player has correctly found the solution to a particularly difficult puzzle, a change in game state from "puzzle unsolved" to "puzzle solved" occurs, and a musical cue enters to signal the player's progress. It is thanks to this music-to-player interaction design that player do not spend a great deal of additional time trying other potential solutions only to find later, in great frustration, that he had the puzzle solved all along, and it had simply been a failure of execution. This scenario demonstrates a clever use of the nondiegetic state-driven interaction scheme.

In one puzzle from *Inside*, the periodic blast of a distant and deadly machine serves as part of the diegetic soundscape. In fact, this blast occurs according to a timed rhythmic interval. If the player character is in an exposed position, not having hidden behind cover when the blast occurs, it spells certain death. The player, therefore, must time his runs from cover according to the machine's rhythmic pulse, which later aligns with the musical score when it enters.

This is a good example of a primary music-to-sound design relationship that blurs the lines of diegesis in order to provide information to the player, all within the context of a nondiegetic state-driven interaction. The scene uses a primary relationship because the sound design is musical: the sound of the machine's pulse occurs in tandem with the game's musical clock. It uses a nondiegetic state-driven scheme because the player's actions cannot directly affect the sound of the machine at all; however, once the player solves the puzzle, triggering a new game state, a new musical layer enters, playing along with the sound of the machine's pulsation, and the music develops according to the same musical clock. Thus, by solving the puzzle, the player has indirectly caused the music to occur in alignment with the diegetic machine sound. ▶(8.9)

Throughout *Inside*, the player encounters a mixture of nondiegetic *state-driven* primary relationships and nondiegetic *player-driven* primary relationships. Music may either be an element attached to an object or a nondiegetic element, albeit one that shares the same general sonic aesthetic as many of the diegetic musical sound effects.

One example of a nondiegetic player-driven primary relationship is when the player approaches a window that reveals a strange, sleeping, child-like creature floating in a room full of water (see Figure 8.9). As the player approaches the window, she is aware from past levels that this creature, if awakened, will stop at nothing to drag the player character down underneath the water to death. While the visuals are thrilling and foment a sense of fear themselves, the terror of this situation is multiplied greatly by the fact that as the player approaches the window, a very subtle, eerie, high-pitched synthesizer timbre creeps into the soundscape. Its entrance and exit are so seamless, as the player either approaches or backs away, respectively, that it seems as if this synthesizer music is always playing near this window (it is indeed possible that this texture is constantly looping in this soundtrack for this moment, automated to fade in and out of audibility depending on the player character's relative distance to the window object). The player can directly interact with this nondiegetic musical element, moving closer to the window or further away, and hearing it grow louder or softer. ▶(8.10)

This moment is also an example of music used to enhance imaginative immersion by instilling a sense of apprehension. Indeed, one has to wonder: does this mean this creature will soon be encountered? Choosing to connect the musical cue to the distance

Figure 8.9.
The creature floats behind the glass. *Inside,* game disc, developed by Playdead (Copenhagen: Playdead ApS, 2016).

Designing Interaction

Figure 8.10
The player must jump over the hole, or perish.

from the window cements an association for the player between the eerie nondiegetic musical texture and the presence of this creature.

This moment, as creepy as it is, is actually a deft setup for a later moment slightly further into the level. After solving a short puzzle, the player has ascended a few stories higher in the game world. The player then proceeds to the right, where he sees a hole in the floor. It quickly becomes clear that the only way forward is to jump over it. As the player draws closer to the hole (see Figure 8.10), the eerie nondiegetic synthesized ambience creeps back into the soundscape, just as it did when approaching the window. The re-emergence of this timbre serves to create fear and anticipation as the player realizes that he is now directly above the creature, and that jumping through the hole would lead to where it is sleeping. This "danger" texture is helpful, for otherwise the relative position of the player to the creature might not be obvious; after all, the player had solved a puzzle in between where he first encountered that sound and his current on-screen location. If there were no such sound here and only the visual element, many players would not realize where this hole leads and the danger it represents. Therefore, the developers chose to provide an extra hint with the music—a nudge to the players to help them realize the significance of their location. ▶(8.11)

The well-considered audio of *Inside* is rife with examples of clever implementation; the author encourages you to study it yourself, paying attention to the different ways the player can interact with the sound throughout the game, and making your own hypotheses as to why certain audio design choices were made.

Conclusion

As we learned in Chapter 4, immersion is a prerequisite for any great gaming experience, and the audio team—composers, sound designers, and audio implementers alike—are directly responsible for much of the success, or failure, of any game. Using tools like Ermi and Mäyrä's SCI model for immersion analysis,[16] we can make observations about how we, as composers, can use music to strengthen the player's experience. The effective composer should consider the various player-music interaction schemes we have explored and use these to create elegant and effective interaction for their soundscapes. Ultimately, good game audio comes from thinking critically about the needs of the game, how using music and sound could affect the player in the processes of the game, and careful application of this knowledge to create an unforgettable audio experience.

Exercises

1. Choose two games to analyze (one first-person shooter, and one puzzle game). Answer the following series of questions for each:
    a. Is the music *diegetic, nondiegetic*, or some combination of the two?
    b. Does the player have *direct* or *indirect* control over the music (is the interaction state-driven or player-driven)?
    c. Is the sonic relationship *secondary, primary*, or *unified*?
    d. Which scheme best describes the player interaction of this game?
    e. What are three instances in which the interactive relationship is effective in strengthening immersion and creating a better gameplay experience? Why?
    f. Name at least one instance in which you feel the game would have benefited from a different interaction scheme.

Notes

1. Karen Collins, *Game Sound: An Introduction to the History and Practice of Video Game Music and Sound Design* (Cambridge, MA: MIT Press, 2008), 4.
2. Ibid.
3. *Call of Duty: WWII*, game disc, developed by Sledgehammer Games (Santa Monica, CA: Activision, 2017).
4. *Horizon Zero Dawn*, game disc, developed by Guerilla Games (San Mateo, CA: Sony Interactive Entertainment, 2017).
5. E.g., *Guitar Hero*, game disc, developed by Harmonix (Mountain View, CA: RedOctane, 2005).
6. *Rez*, game disc, developed by United Game Artists (Tokyo: Sega Games, 2001).
7. Perhaps so-named because the camera and the player's avatar seem to be attached to a track, a *rail game* is a game in which the character's route of movement is predetermined. In a rail *shooter*, the player's main goal is to shoot the enemies and obstacles as they appear on-screen.
8. *Electroplankton*, game cartridge, developed by indieszero (Kyoto: Nintendo, 2005).
9. *Portal 2*, game disc, developed by Valve (Bellevue, WA: Valve, 2011). The gameplay snippets were originally found on the "portal2soundvideos" YouTube channel from a useful compilation entitled "Portal 2: Interactive Music," provided by portal2soundvideos: portal2soundvideos, director, Portal 2: Interactive Music, YouTube, January 4, 2012, www.youtube.com/watch?v=ursIj59J6RU&t=324s.
10. *The Legend of Zelda: Breath of the Wild*, game cartridge, developed by Nintendo EPD (Kyoto: Nintendo, 2017).
11. *Bendy and the Ink Machine: Chapter Two: The Old Song*, download, developed by theMeatly Games (Ottawa, ON: theMeatly Games, 2017).
12. *Thumper*, download, developed by Drool (Providence, RI: Drool, 2016).

13. *Uurnog Uurnlimited*, download, developed by Nifflas Games (Stockholm: Raw Fury, 2017).
14. Niklas Nygren (a.k.a. "Nifflas" of Nifflas Games), interview with author, December 3, 2017.
15. *Inside*, game disc, developed by Playdead (Copenhagen: Playdead ApS, 2016).
16. Laura Ermi and Frans Mäyrä, "Fundamental Components of the Gameplay Experience: Analyzing Immersion," *Changing Views: Worlds in Play—Selected Papers of the 2005 Digital Games Research Association's Second International Conference* (British Columbia, Canada: Vancouver, June 16–20, 2005).

PART III
Advanced Reactive Music Concepts

9

Emergence: Music's Role in Player Experimentation

Introduction: What Is Emergence?

In many video games, how to beat a level or accomplish a task tends to be rather clear or obvious. For example, the player might need to eliminate a band of enemies who arrive in sequential order by using the assault rifle with which he or she began the level, or he or she might be tasked to climb a series of platforms to reach the top using the "jump" button. In either case, a clear path has been laid out, and the player will most likely attempt to follow this path.

Conversely, some games are designed to encourage the player to find new ways of beating levels—or at the very least, to give him or her the freedom to do so. These games are often built systematically, meaning the various systems and objects in the game interact not only with the player, but with each other as well. This creates a world full of possibilities, much like the real world—a world that players can experiment and interact with, finding their own solutions to problems, according to their own personalities. Many of these games are *sandbox games*, which place little-to-no limitation on the player's activities, allowing the player to explore the world and interact with it in whichever way he or she deems fit. The term originates, of course, from the concept of playing in a sandbox: within its walls, there are no artificial constructs or rules for how to use imagination to have a good time.

In an environment that is pre-programmed to encourage such variability, new behaviors often emerge, be they within programmed systems or between players in multi-player games. In her discussion of sandbox games, author and game designer Briar Lee Mitchell briefly touches upon this concept of *emergence*: "To allow you to act freely, a lot of variety in game mechanics is built into these [sandbox] games. A term for that is emergent gameplay: basically, making the world interactive enough that the player can do things the designers hadn't originally intended."[1] In other words, when a player executes a new behavior outside the clear purpose or method of gameplay, that behavior is *emergent*.

Since emergent gameplay means that each player will experience a world in a different way, the music system in such a game should also be capable of reacting to this variability. All players will experience the soundtrack differently, according to their play styles. Thus, just as the developer strives to make the sandbox game fun, open-ended, and personalized, the composer should work hard to design music systems that have the ability to react to any possible situation. In this chapter, we will go over some different examples of emergent gameplay and its relationship to game scoring. We will learn how to design complex music systems that can react to practically any player decision. We will also see how we can encourage emergent behavior in a player who is interacting with our music.

What Is Systematic Game Design?

Systematic games are designed using a hierarchical tree of *systems* that interact with each other as well as with the player. *Systematic gameplay* tends to be found in many open-world, sandbox games, giving the player a great deal of freedom and encouraging exploration and unique, personality-driven problem-solving. Because of the many possible interactions between game objects, the gameplay is variable and unpredictable. Players can discover altogether new ways of interacting with the world by manipulating objects and their relationships. To quote seasoned game designer, Mike Sellers:

> A system is made of parts that interact to form a purposeful whole. . . . Each "part" in a system can be thought of as an object in software terms, with its own state (defined by attributes) and behavior. Importantly, this definition is recursive: each part is itself a whole, a sub-system, with its own internal workings: a part's attributes are themselves objects with their own states, attributes, and behaviors.[2]

Each object in a systematic game has a set of "ears" (the input) that listen for messages from other systems, and a "voice" (the output), which is constantly announcing its current state. Each object is constantly listening for an output message from either the player, another machine, or an NPC, which aligns with its input. If and when an input and output do align, this triggers a certain *rule,* as designed by the developers, to be followed. This type of variable interaction between objects opens up a number of possibilities for players to manipulate the world around them in order to solve problems. While the most obvious path to defeating a group of enemies might be to simply battle them head-on, a systematic game might encourage emergence by giving the player the option to instead win a battle by turning enemies against each other and allowing them to battle among themselves instead; or, perhaps, by using a torch to light a wooden hut on fire, causing a distraction that allows battle to be avoided altogether; and so on. In *Horizon Zero Dawn* (2017),[3] for example, the monstrous robotic dinosaurs will enter battle with the player whenever they are triggered to do so; however, they will also interact with other systems in the game, battling NPCs or even other machines. The player, therefore, has the freedom of choice to take on a group of enemies head-on by using a weapon—or, perhaps, to merely trigger a battle between the enemy NPCs and the machines, thereby distracting them and encouraging them to kill each other while passing through the level undetected.

Systematic game design encourages emergent gameplay. The more intricately the different systems in a game can interact, the more possibilities exist for the player to find unique ways of beating levels and solving problems without taking the most "obvious" path.

Systematic Music Design

We will use the term *systematic music design* to refer to music designs that encourage emergent behavior when players interact with our music. Just as game developers use systematic game design to plan for emergence and encourage players to interact with their games, systematic music design is the concept that musical elements are also constantly listening for messages from the other systems in the game. Thus, they can interact with other game objects, depending on the rules established by the composer and developer. To simplify further, the music system is yet another system with inputs and outputs that can interact with the game as a whole.

No Man's Sky (2016),[4] by Hello Games, has over 18 million explorable planets, a feat made possible by using a real time generation system to create planets (we will touch upon how this affects the music in Chapter 10). Because the game's landscapes are built anew each time, audio designer Paul Weir had to begin his audio-creation process without knowing what each planet might look like, which objects might be part of the planet's environment, where the objects might be located, or what creatures might inhabit the world. Similarly, players may encounter many different types of terrains when exploring a single world, so their actions will also be somewhat unpredictable and dependent on the terrain.

In order to account for this variability, Weir decided to take a system-driven approach to designing the audio system. In his 2017 GDC talk, "The Sound of *No Man's Sky*," he pointed out that simple event-driven audio would not suffice. It is for this reason that he took a system-driven, multilayered approach.[5] *No Man's Sky*'s audio system is composed of numerous events and behaviors that are grouped in a hierarchy, each level containing a series of smaller commands within it as messages provide more detail. These objects are constantly listening for messages from outside sources. When the output of a game object aligns with the input of an audio object, this triggers the audio to react according to the established rules.

This hierarchical approach to systems design is quite effective. Using *Wwise*, Weir built a series of systems using "states" and "switches," meaning he could create a hierarchy of instructions for the audio system to abide by in order to organize it in response to the data sent to the audio system. He created four different states to coincide with the four overarching "biomes" that can be encountered in the game. Each of these biomes has its own unique atmosphere or environment, which means its creatures, buildings, plants, terrain, and even weather will fall within a certain broader category. Once this broader category has been determined, subcategories can subsequently be determined as the planet is generated, providing more information to the audio system about what types of creatures, plants, terrains, and scenarios might occur in the instance of this particular planet. Since the player can only ever be in one specific biome at any given time, Weir used states to control the audio. Similarly to how the procedural generation system limits what visual elements exist in a certain biome, Weir's audio system limits which sounds are likely to occur in each biome.

The music system itself was also built using a top-down approach: the larger, overarching categories of the score were defined first, thus allowing the composition team to get more and more detailed, building smaller possibility trees within the larger categories. To create the score, experimental rock band 65daysofstatic took to the recording studio to lay down an album of music for the game. While their original composition was done linearly, much like a traditional score, they wrote their music with the knowledge that their album would be divided into many pieces and put back together for *No Man's Sky*. Thus, they returned to the studio after recording the "core" album to create new guitar riffs and drum loops and remix all of the original tracks for use by the game's procedural music engine.

In the end, the band compiled over 2,500 samples specifically created for the purpose of being utilized and recombined in the game.[6] The game itself has no real narrative structure and is instead geared toward providing each player with a unique playing experience. The development team wanted the score to likewise provide each player with a highly personalized musical soundtrack. The music is organized into several overarching categories: "Planet," "Space," "Combat," "Map," and a few others for special situations. Within these categories, there exist roughly 50 different soundscapes. Of course, the music is also tied to the biome system. Since each planet is unique, musical elements are tied to many of these possible characteristics of potential planets, and the different states and switches controlling these are triggered in *Wwise*. The team then put together groups of sounds into instruments that would play these sounds, depending on a number of factors being fed by the game engine. The music system reacts to output from the engine, giving the player a great deal of influence over the resulting score. For example, the score reacts to "space" and "direction," meaning that the music will react based on which direction the player is facing and what is in the player's field of vision. The score also changes depending on whether or not a player is moving and the player's proximity to specific places or landmarks. Conversely, if the player has been walking for quite some time without any "interesting," state-changing event taking place, the music will begin to rise in intensity slightly to maintain the perceived level of interest in what's happening on-screen.

For a vast, open-world game that encourages emergent gameplay, a complex music design is often necessary to account for the many directions in which a player can go. *No Man's Sky* offers an interesting look at a successfully implemented systematic music design. It takes a top-down approach to systems design, categorizing its musical possibilities hierarchically. Let's explore this approach in greater detail by looking at another example of a sandbox game: *Just Cause 4* (2018), by Avalanche Studios.[7]

Vertical Layering and Horizontal Resequencing in Advanced Systematic Music Design

The two fundamental game scoring techniques of *horizontal resequencing* and *vertical layering* yield vast musical possibilities when combined within a dynamic music system. One game that uses this combination effectively and liberally is *Just Cause 4*, a triple-A, open-world, sandbox action/adventure game based in the fictional South American country of Solís. The *Just Cause 4* music team, spearheaded by technical sound designer

Ronny Mraz and composer Zach Abramson, created a complex modular design using a combined vertical and horizontal structure to enable a detailed reactive score.

As we discussed in Chapter 2, horizontal resequencing is a method of designing a game score that focuses on the "left-to-right" development of a composition and how it changes dynamically based on game data and player action. It allows the creation of long-form musical structures which can account for multiple timeline possibilities, transitioning smoothly from one segment of music to the next. Vertical layering, on the other hand, is less focused on the horizontal form of a composition; rather, it is centered around instrumental stem layers being added and subtracted over a base layer defining the underlying harmony or formal structure.

These two techniques are some of the most basic tools of any game composer's skill set. While vertical layering is used more commonly, most scores will use a combination of these two techniques, depending on what best serves each circumstance. With a creatively implemented system, the combination of horizontal resequencing and vertical layering can allow a wide range of possibilities when creating a reactive score.

The biggest strength of the vertical approach is that each layer in a composition can be easily attached to incoming game data. A composition's orchestration may be set up to build in complexity based upon conditions like combat intensity, proximity to specific locations, or proximity to characters, to name a few. Each layer can be affected individually, meaning the composer has a great degree of control over how the score reacts to the player's decisions. The horizontal approach has its own strengths; what it lacks in terms of dynamic instrumentation, it makes up for with transitional capability. Indeed, Abramson points out just how important a creative imagining of transitions is in the *Just Cause 4* music system:

> From a composition standpoint, our transitioning concepts had to do with making sure that changes between different types or sections of music never felt abrupt or jarring, and that each transition was effectively supporting changes in the gameplay experience—all while making sure that there was cohesion between each bit of music. We used stingers to ramp up and ramp down intensity and used common tonalities and key centers to make music feel as though it "cadenced" to different music states as it transitioned. Lastly, we were very attentive to the loudness and the frequency range of the music so that transitions felt seamless against changes in the overall sound design and mix.[8]

Combining vertical and horizontal techniques offers a far more complex array of functionality for emergent musical structures. Vertical layers can work with individual transitions, thereby having the freedom to change with the horizontal score, or to act on their own. For example, the top layer of the instrumentation can remain, while the bottom layers transition to an entirely new piece of music. All in all, this combined system is a powerful one, but it must be implemented with a great degree of organization and thoughtfulness. Through their own careful planning, the *Just Cause 4* audio team did just that. Their score's functional success offers a glimpse into how a combined horizontal-vertical structure can be applied to a game music system so that each player can experience the score differently.

Advanced Reactive Music Concepts

Creating Structure: Going Modular

Implementing an advanced, combined music design with the power to score emergent gameplay is best achieved using a *modular system*. In a modular system, the score is mapped out according to a series of musical "modules," each one based on a piece of the larger compositional structure. Each module complements a specific situation in the game and can differ in length, depending on what may be needed for each circumstance.[9] One advantage of utilizing a modular system structure is that modules can smoothly transition between each other. By combining horizontal resequencing and vertical layering, each module can have various sublayers that enter and exit vertically, while the music can develop horizontally between different modules as the gameplay itself develops and changes.

Much like the system-driven audio in *No Man's Sky*, this is a top-down approach, and successful implementation requires excellent organization from the beginning of the project. The first step is to decide upon a series of upper-level, overarching ideas or sections. The example from *Just Cause 4* as outlined in Figure 9.1 shows how a vast, complex design is broken down into a few key structural elements at the beginning of the process.

Within each of these overarching modules, smaller modules are created and set up to transition to each other, depending on the messages received at any given time from the game engine. Within each of these smaller modules, sub-layers exist which can respond to even more granular levels of game parameters. Such a structure can become very complex—indeed, the composer may need to account for dozens, or even hundreds of possibilities with distinct, predetermined musical reactions. Figure 9.2 shows an example of this complexity after all of the details were added to *Just Cause 4*'s "Strike Music" structure.[10]

As complex as this diagram is, it outlines only a single type of combat in *Just Cause 4*, which features multiple combat types, as well as many other musical structures. Likewise, a hierarchical music matrix diagram was created to organize and visualize the musical possibilities of each of these other structures. Putting it all together forms a large possibility tree. This is why it is important to take this into consideration before creating the system.

Figure 9.1.
Example diagram of upper-level hierarchical system design in *Just Cause 4*.
Line drawing by Ronny Mraz and Zach Abramson.

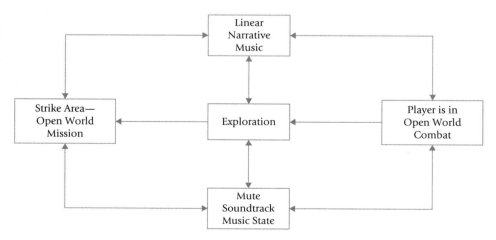

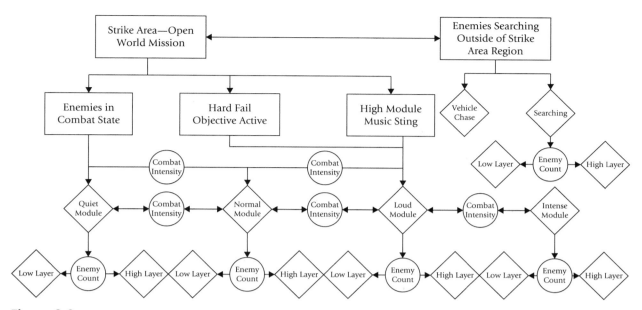

Figure 9.2.
Example diagram of upper-level hierarchical system design with more detail in *Just Cause 4*.
Line drawing by Ronny Mraz and Zach Abramson.

Exploring *Just Cause 4*'s Modular Horizontal/Vertical Music Design

Just Cause 4's overarching game design is focused around four main pillars: *Verticality, Variety, Destruction*, and *Living World*.

1. *Verticality*: The game world must function freely and cohesively in the 3D space. The game offers a great deal of travel methods, most notably aerial navigation through the use of a wingsuit and parachute, so the world must be fully cohesive from the sky down to the ground.
2. *Variety*: The game features an eclectic environment that offers many different unique gameplay opportunities to the player.
3. *Destruction*: Most objects in the game can be destroyed in grand or surprising ways that are deeply satisfying to the player. This is a core aspect of gameplay in the whole *Just Cause* series.
4. *Living World*: The game should feel like it has a living, breathing atmosphere. Since *Just Cause 4* is a sandbox, open-world game, the dynamism of the entire world is paramount to a fun playing experience for the player.

Mraz, who worked hand in hand with Abramson on the score's structure, stresses that the Avalanche audio team determined early on that the musical score itself should be built to adhere to and strengthen these four pillars of the game design.[11] Therefore, the score was made very reactive to the game's Verticality, complementing the player's experience and differing depending on whether gameplay is occurring in the sky or on the ground. The music also complements the Variety by a diverse range of orchestration, dynamics, and overall feel across types of gameplay and game-world locations. The music easily embodies the bombastic Destruction of the game as a powerful, cinematic action score. Finally, the music's detailed reactivity works to enhance the Living World by using complex reactivity to let the score itself react to stimuli as if it were itself alive.

To best handle the complex needs set by the series of goals outlined in these four game pillars, the *Just Cause 4* music team decided to use a modular music system design. They began by designing, in their words, "macro music states," another good example of the "top-down" approach. These "macro" states encompass the many "micro music states." Thus, this hierarchy allows their system to function by first determining the larger states, then by checking for increasingly detailed sets of information as it "moves down" the hierarchical ladder. Within the Exploration Music macro, for example, the system first checks whether the player is in a Flying state or a Ground Exploration state. If the player is flying, the music enhances the feeling of movement experienced through flight with repetitive rhythmic motors. If the player is in a Ground Exploration state, then the music is more ambient, blending with the environment and allowing the soundscape of the surroundings to dominate. Then, the system progresses into the micro states. If the player character is on the ground, is the character in Danger, in a Special Location, or merely exploring the biome's environment? If the player character is in the air, is the character using a wingsuit or parachute; and is the character in Danger or Near the Ground? The confluence of all of these determinations produces a unique musical result. With this level of details, the score is custom-tailored to each player's individual playthrough actions.

As Mraz further elaborates:

> In combat, for instance, a micro music state could be "Combat Low Intensity" or "Enemies are searching for players in a vehicle chase." In exploration, micro states could be: ground or air exploration, or additionally, "Player is flying close to the ground while surrounded by enemies." A macro music state's micro music states could be as general or granular as their situation calls for, but each micro music state is specific to its corresponding macro music state.[12]

Biomes also play an important role in the resultant macro states. The player may be in a Rainforest Flying Exploration state. At the micro level, the vertical layers change based on micro state changes. When the player is using a parachute—a slower, safer way of descending than the wingsuit—the music remains slow-paced, based on a quarter note rhythmic pulse, leaving more temporal space. If, however, the player switches to using the wingsuit—a much more dangerous and fast-paced method of travel—the percussion layer switches to an eighth note–based rhythm, increasing the level of musical intensity. The music also reacts to the player's proximity to certain objects. For example, if the player is flying fast in the wingsuit near the ground, a "near ground layer" would enter with a sixteenth note rhythmic pattern, heightening the intensity still further. Our first Chapter 9 companion site example, generously provided by Mraz and Abramson, demonstrates two functions of their music system while the player is in the wingsuit: first, how the music reacts at the micro level to the danger, depending on the player's proximity to the ground, and second, how the music also changes at the macro level based on the current biome. ▶(9.1)

When the macro state changes from Exploration to Combat, the combat micro states also begin to influence the music. For example, if the player is circling the edge of a combat area, a Dangerous Layer fades in, slow and anticipatory in nature. If, however,

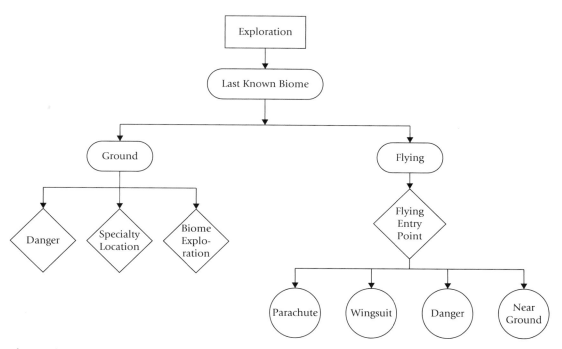

Figure 9.3.
Example diagram of music implementation in *Just Cause 4*'s "exploration" state.
Line drawing by Ronny Mraz and Zach Abramson.

the player nears the frontline of the battle, the Player Near Frontline module would be triggered, and the music would become more sparse, leaving space for the cacophonic diegetic battle sound effects that would be occupying the majority of the soundscape.

In addition to the macro music states, instrumentation is yet another factor accounted for by the *Just Cause 4* music system. The game's *FMOD* session contains 50 constantly monitored game parameters, all of which have different effects on the music.[13] For example, which biome the player is in has a direct effect on the instrumentation of the currently playing score. Our second companion site example illustrates the transition from exploration on foot, to exploration in the wingsuit, as well as some subtle instrumental changes when crossing over between biomes. ▶(9.2). If the player is entering an Open World macro state, the engine checks the current biome before proceeding into the corresponding gameplay area. The instrumental and/or rhythmic layers will vary a great deal from biome to biome. Figure 9.3 demonstrates the flowchart detailing how the music was implemented for the Exploration game state.

Implementing a Complex Modular System

When designing a modular system, it is important to consider the technical possibilities and limitations of the software system it will be developed on. Middleware software allows the composer to "program" the music without having to write any actual code. As we will see in Chapter 15, each middleware software offers its own solutions to common video game music implementation problems. While *FMOD* one of the leading middleware softwares available today by Firelight Technologies, provides what is arguably the most user-friendly audio workflow, as it is based upon a linear timeline view, other middlewares like *Wwise* and *Elias* each have distinct advantages, too.

In the case of *Just Cause 4*, *FMOD* provided a simple and intuitive platform with which to integrate the combined horizontal and vertical system. Because of its unique timeline view, it was possible for the audio team to place all of their macro states within a single "event," an audio asset in *FMOD* that can be used and manipulated by the game engine. Each macro state was placed on its own track (visually, much like a track in a DAW). Then, the appropriate micro states were added into each event as "subtracks," tracks that live within the hierarchy of a normal track. This workflow allowed the team to easily build a session that corresponded to their music design both visually and functionally.

The use of middleware allowed the *Just Cause 4* team to build upon their hierarchical music systems without the need to create new workflow solutions to handle these expansions. Having music completely controlled by the functions of the middleware also allowed the audio programmers to focus on other elements of the audio system.

As convenient as middleware implementation is, though, there may drawbacks to using a middleware solution when detailed reactivity is a core system goal. Any preconceived software solution comes with its own limitations to how it can be used and what can be accomplished. Because of these constraints, a middleware-reliant system tends not to be as scalable because a new function cannot be simply built as needed. Debugging errors also tends to be more difficult because the audio programmer often does not have a direct window into the audio system's code. Ultimately, in many cases, it is best to use a combination of custom tools alongside middleware solutions, for the best of both worlds. For the new indie composer without programming experience, however, finding ways to be creative within the constraints of established middlewares will often be the most efficient solution.

All in all, *Just Cause 4* is an example of a game with advanced reactivity that accounts for a large-scale sandbox game with numerous examples of encouraged emergence. The music system designed by Ronny Mraz, Zach Abramson, and the rest of the Avalanche team can account for a truly impressive array of possibilities, providing the player with a custom musical score tailored to the epic action experience provided by the gameplay.

Designed Musical Emergence in *Sound Sky*

While the sandbox games *No Man's Sky* and *Just Cause 4* are excellent examples of effective systematic music design implementation in their genre, there are other styles of games that also encourage emergent gameplay, albeit on a smaller scale. One such example is *Sound Sky* (2019),[14] an indie music performance and music creation game by developer Highkey Games. The game's core mechanic revolves around the player executing rhythms along with the electro-tropical score by composer Craig Barnes. Barnes, who played an integral part in the game's music design process, created the music puzzles in all of the levels so that they fit with his musical tracks. The gameplay blurs the lines of diegesis as the player's finger-presses trigger drum samples that "complete" the drum track to each song.

As shown in Figure 9.4, the green, blue, and pink circles indicate the rhythm to players, who must use their fingers to tap the corresponding buttons at the precise moment the beat arrives to each. As the sample sounds, players "complete" the drum track,

Figure 9.4.
Example *Sound Sky* UI/gameplay screen.
Screen capture of *Sound Sky* by composer Craig Barnes, provided with permission.

thereby playing an active role in the performance of the music. This gameplay mechanic has been proven to be effective by numerous music games before, most notably the *Guitar Hero* series,[15] in which the player must time button presses on a guitar-shaped controller with the song part scrolling on the screen.

A particular feature note in *Sound Sky* is its music creation system, which gives players a striking degree of control over the music, both allowing them to create their own levels and encouraging them to become creative, taking the music design into their own hands. As Barnes states,

> We [the Sound Sky music team] wanted to give [the player] the ability to create their own levels, so we designed an "entry-level" music making tool that tricks [the player] into writing a song. [The player collects loops] from the music I've written, but then [is] able to mix and match them to create [their] own remixed tracks [. . . and their] own sequenced rhythms.[16]

Allowing players to create their own tracks as an additional gameplay mechanic is notable as it offers a new way for the player to interact with the game other than the game's core rhythm-execution mechanic. Algorithmic music requires the composer to essentially also be the designer of a formal rule set that determines how the computer may generate music in real time based on game input data. Designing such a rule set takes a great deal of preplanning and organization. In the case of *Sound Sky*, the development team had to establish a rule set that would grant the players the power to create their own versions of the game's music, thereby inspiring a feeling of musical liberty and encouraging creativity, while also governing that process enough to guarantee a certain consistent level of quality in the achieved musical result. Successfully implementing this feature was undoubtedly no easy task, conceptually or technically.

However, Barnes's involvement in the game design process from a very early stage allowed him and the development team the time they needed to create a system with the functionality that this feature required. Barnes divided his musical stems into smaller horizontal segments and vertical layers, which could later be combined into user-created songs. These segments and layers include chords, melodies, two different keys, tempi, *feels* (i.e., "straight" or "swing"), and chord progressions. Each of the segments has looping capability. To ensure musical cohesion, all of the loops were produced in the keys of either C major or A minor. All of the loops are also based on particular harmonic progressions so that they can easily be transitioned between without the result seeming musically out of place (and therefore non-immersive).

Thus, the designed emergence in *Sound Sky* comes with some limitations for the player so that the musical result always sounds appropriate. When the player picks a loop, the possibilities for which loop might come next are limited to those that had been predetermined by Barnes. Barnes accomplished this logic design by establishing rules that limit the player's "compositional" possibilities—for example, if the first loop chosen is in C major, the song is established as being in C major, meaning loops in A minor become unavailable from this point on. If the first loop's feel is swing, then all of the following loops will be swing. Once the player has chosen a specific bassline outlining a particular harmonic progression, it would no longer be possible to pick parts following a different harmonic progression, which would obviously clash with the bassline.

To encourage emergent behavior, a "random" button was added to the game's interface. When it is clicked, it inserts a randomly chosen horizontal segment as the next section of music. This feature provides the player with, to borrow Barnes's words, "happy accidents." This randomization, though, is still limited to occurring within Barnes's rule sets. With regard to melodic content, chromatic melodies or dense diatonic melodies would have been limiting to use because they have greater potential to clash unacceptably with different harmonies. For this reason, the melodies in *Sound Sky* are generally based on the pentatonic scale. The open intervallic structure of the pentatonic scale allows melodies to be flexibly realized over many possible harmonies.

The game also includes various tempo changes throughout the gameplay. Due to the limitations of *Unity*'s built-in audio engine, changing the tempo of a loop results in its pitch changing as well, potentially creating harmonic dissonance between loops originally recorded at different tempos. Barnes solved this problem by bouncing out all of the audio at a single tempo, 106 beats per minute (bpm). Whenever the loops change tempo in the game engine, that change remains consistent across all loops, negating any potential variation in pitch between loops. To again quote Barnes:

> If we're using the same loops for different kinds of songs in different tempos, we want to ensure that all loops change both tempo and key at the same rate. Say we had a melody bounced out in C major at 106 bpm and a bassline bounced out in C major at 140 bpm. If we wanted to create a song at 140 bpm and use both this melody and this bassline, increasing the tempo of the 106 bpm melody up to 140

bpm would cause it to no longer be in C major—but the bassline would still be in C major. So all loops are bounced out at the same tempo to get around this.[17]

All in all, Barnes was able to create a music system that encourages the player to experiment and take a central role in the resequencing of the game's soundtrack. This functionality builds a uniquely personal connection between the player and the score.

Organic Emergence and *Guild Wars*

At this point, we should make a distinction between two different types of ludic emergence. In some cases, emergent behavior is encouraged by the game itself: *designed emergence.* For example, in a (real) sandbox, children may be given certain toys that encourage them to play with the sand in specific ways, though the ultimate result is unpredictable. However, emergent behavior can also occur when players create or discover completely new ways of interacting with the game world, ways the developers had not conceived. When this occurs, it is called *organic emergence.* In some cases, emergence may result from players reacting to a programming bug, or even purposefully exploiting the bug together once it has been discovered. In his book *Basics of Game Design,* Michael E. Moore interviews game designer Chris Taylor (*Total Annihilation* [1997],[18] *Dungeon Siege* [2002]),[19] who says ". . . as we [game designers] have gotten more sophisticated in the craft of game design, we kind of know how to encourage that kind of [emergent] gameplay."[20]

Indie composer Lena Raine may be best-known for her soundtrack to indie darling *Celeste* (2018);[21] however, she also worked on another well-known game called *Guild Wars 2* (2012),[22] a *massive multiplayer online role-playing game* (MMORPG). In *Guild Wars 2*, players can obtain musical instruments and control how their in-game avatars play them. Raine points out that this feature eventually led to a number of emergent behaviors:

> My very first music project on that game [*Guild Wars 2*] was making a bell choir mini-game, kind of like [a] *Guitar Hero*–style thing. You [the player] have a full octave of bells to play. It was a structured mini-game that was like, "here's a bunch of pieces I've written and you're gonna play them." Then, we gave the bells to players to let them play stuff in the actual game. That created this huge wave of in-game instrumentalists who were taking their bells, making music with others, and even performing for people. People started forming bands and playing together in the world. We were basically just giving them the tools to make their own music in the game.[23]

Guild Wars 2 is a powerful example of how giving players certain freedoms can lead to completely organic and unexpected emergent behaviors. As of this writing, there are a number of *Guild Wars 2* band performances available for viewing on the internet. In these cases, a large group of players "got together" in the game to perform their favorite real-world music using the in-game instruments.

While developers by definition are not able to predict when or how organic emergence will occur, the savvy composer can, at the least, consider creative ways to expand the player's freedom to interact with the game world (and, perhaps, with other players) on the musical level.

Conclusion

Emergence is an incredibly important concept to understand. Indeed, as systematic game design inevitably continues to develop as a craft in tandem with the technological advancement of gaming hardware, we will find ourselves writing music for worlds that allow more and more unpredictable player behavior. We must first understand how these worlds have been designed and how they function so that we can craft audio systems that can musically react to any of the many possibilities which might occur during gameplay.

Just Cause 4 offers a glimpse into some of the creative possibilities when using a combination of horizontal and vertical sequencing to create a music system which can account for numerous possibilities that can occur in an emergent game. Ronny Mraz, Zach Abramson, and the rest of the Avalanche audio team were able to create a robust music system to match the scale of the game itself. Because of the incredibly granular gameplay possibilities, the music had to reflect this dynamic landscape with its own high-level musical IQ. By using horizontal and vertical sequencing in combination with detailed game parameters and triggers from the game, they were able to design an intricate music matrix that could respond to almost any situation in the game.

Sound Sky illustrates how emergence can actually be encouraged from the musical standpoint, giving players the freedom to interact with the score in their own ways. *Guild Wars 2*, on the other hand, is a good example of how giving players certain functions in a game might lead to consequences you would never expect.

All in all, you should always try to keep an open mind during the music design process. If you are working on a systematic game with great potential for emergent behavior, think through the game as if you were playing it (of course, play it yourself if possible), and consider how you might be able to create a systematic music design to account for any kind of emergent behavior.

Exercises

1. Research three different sandbox games and their music systems. Choose one of your favorite examples.
2. Spend an hour either playing the game, or watching its gameplay. As you do so, take note of the many different types of scenarios that may or may not occur during gameplay.
3. Begin designing a top-down system-driven hierarchy using modules. First, create three to five macro states that will control the upper-level decision-making of the music system.
4. Next, begin designing micro states. Get as detailed as possible.
5. Draft a full hierarchical diagram that displays your music design and how it accounts for all possible gameplay scenarios.

Notes

1. Briar Lee Mitchell, *Game Design Essentials* (Hoboken, NJ: Sybex, 2012), 6.
2. Mike Sellers, "Systems, Game Systems, and Systemic Games," *Gamasutra* (blog), May 18, 2015, www.gamasutra.com/blogs/MikeSellers/20150518/243708/Systems_Game_Systems_and_Systemic_Games.php.
3. *Horizon Zero Dawn*, game disc, developed by Guerilla Games (San Mateo, CA: Sony Interactive Entertainment, 2017).
4. *No Man's Sky*, game disc, developed by Hello Games (Guildford, UK: Hello Games, 2018).
5. Paul Weir, "The Sound of 'No Man's Sky,'" Game Developers Conference 2017 (convention lecture), Moscone Center, San Francisco, CA, March 3, 2017.
6. Ibid.
7. *Just Cause 4*, game disc, developed by Avalanche Studios (Tokyo: Square Enix, 2018).
8. Zach Abramson and Ronny Mraz (*Just Cause 4* music team), interview with author, April 1, 2019.
9. In *Just Cause 4*, for example, musical modules are generally between 30 and 40 seconds long.
10. *Strike missions* are combat missions that can be failed if an objective is not completed, as opposed to general open-world combat, which is free play without any real consequence.
11. Zach Abramson and Ronny Mraz, interview with author, April 1, 2019.
12. Ibid.
13. Ibid.
14. *Sound Sky*, mobile, developed by Highkey Games (Cupertino, CA: Highkey Games, 2019).
15. E.g., *Guitar Hero*, game disc, developed by Harmonix (Mountain View, CA: RedOctane, 2005).
16. Craig Barnes (composer of *Sound Sky*), interview with author, February 26, 2019.
17. Ibid.
18. *Total Annihilation*, game disc, developed by Cavedog Entertainment (New York: GT Interactive Software, 1997).
19. *Dungeon Siege*, game disc, developed by Gas Powered Games (Redmond, WA: Microsoft, 2002).
20. Michael Moore, *Basics of Game Design* (Natick, MA: A K Peters/CRC Press, 2011), 350.
21. *Celeste*, download, developed by Matt Makes Games (Vancouver, BC: Matt Makes Games, 2018).
22. *Guild Wars 2*, download/online, developed by ArenaNet (Pangyo, South Korea: NCSoft, 2012).
23. Lena Raine (composer of *Guild Wars 2* and *Celeste*), interview with author, March 1, 2019.

10

Proceduralism
Using Game Data to Create a Real Time Score

Introduction

In his talk at GSC[1] 2018, music industry veteran Thomas Dolby brought up a pertinent issue with current audio formats: audio is treated as discrete, isolated units (e.g., left channel, right channel; guitar, bass, drums) that is manipulated in a DAW and then rendered (or bounced) out into a mostly permanent form. He argued that this workflow is contrary to the essential flow of nature. Nature, he says, does not act in discrete chunks, but rather with *collective intelligence*. Thus, it is not the individual units that are significant, but rather their behavior as parts of the whole. He proposes, then, a new paradigm of audio format with a great deal more metadata attached so that the audio files themselves can communicate with each other, serving as a part of a collective audio intelligence. In this hypothetical format, the audio files could adapt and evolve based on the context of their environment. Instead of being locked into fixed processing and automation, audio could react and change based on algorithms and real time data.[2] While his proposed audio format has yet to be invented, current game audio creators are already working around the sonic limitations of pre-rendered audio formats using *procedural audio*.

Procedural (or *generative*) *audio* describes audio created algorithmically in real time based upon input (incoming) data, or data received from an outside source. Composers and sound designers who create procedural music and audio are therefore also quasi-programmers—as opposed to relying on the traditional pre-rendering of audio files, they create a set of rules (that is, the algorithm) which govern how the computer will create and perform the resulting sounds, depending on what is happening in the game. When audio is procedural, it is being realized in *real time;* everything from musical form to instrumental timbres, from when instruments play to where those instruments positionally belong in the soundscape, can be written into the algorithm so that it is determined only a split-second before the music is created by the processor. The result is a fully reactive score—one composed on the spot as a direct result of what is happening on-screen.

In this chapter, we will use a series of case studies to examine the various ways in which procedural techniques have been used to generate real time scores in games. We will also discuss the relative strengths and weaknesses of procedural scoring against traditional audio composition and mixing conceptions, and how and when those attributes might affect or influence the use of one style or the other. Finally, we will talk about some ways in which you can apply procedural scoring to your own game projects.

What Is Procedural Audio?

The concept of proceduralism is not limited to audio. In fact, procedural generation is used for other purposes as well. One good example is its usage in the creation of large, complex game worlds. It can be particularly useful for creating textures and environments in open-world, exploration-based games with scales far outreaching what developers could realistically hand craft. For example, creating the 18 million individual, explorable planets in *No Man's Sky* one by one would be impossible to achieve in a single lifetime, particularly for a small development team like Hello Games. Thus, to accomplish this staggering feat, they implemented a procedural system to generate all of the environments in real time using preset parameters controlling a finite number of assets in their algorithmic system.

There is one important distinction we must understand: even if audio is generated in real time, this alone does not make that audio procedural; the audio must be generated algorithmically, meaning that it is created as a response to input data. Developments in the gameplay—which, as we have discussed before, is necessarily nonlinear and interactive—are fed to the musical algorithm, which in turn effects (and affects) the music. Indeed, in the talk he gave at GDC[3] 2017, Hello Games audio director Paul Weir affirmed that, though any form of real time audio synthesis is commonly referred to as "procedural audio" simply because the sounds and pitches happen to be generated in that moment,[4] audio merely being generated in real time is not enough for it to be classified as procedural; rather, in order to truly be procedural, the audio must be generated in response to real time game data as well. True procedural audio, a collaboration between the composer, the computer, and the game itself, creates an exceptionally unique score for each gameplay.[5] This is an important and logical distinction.

Aside from the aforementioned uniqueness of each musical playthrough, procedurally generated music has a number of other notable differences from using a pre-rendered score. Often, these differences are advantageous. Situations requiring dynamic tempo changes, for instance, easily expose the limitations of pre-rendered audio when compared to procedural audio. In order to change the tempo of a cue in response to the state of the gameplay, a pre-rendered track would either have to be time-stretched in real time using software, or numerous versions of that track at different tempos would need to be bounced and layered, ready to be triggered at a moment's notice.

Neither of these solutions is ideal. Time-stretching an audio file, particularly if it is a larger file like a minutes-long music loop, is an imperfect process, often producing audible glitches and artifacts in the resultant audio—even when processing is done *offline*.[6] The alternative is even more clunky: having a variety of pre-rendered loops at all possible tempos means that the amount of audio assets must increase exponentially, since each track now requires multiple concurrent versions; and not only this, but these loops must be kept in constant proportional synchronization, such that moving between them creates a seamless musical experience. Therefore, under the traditional pre-rendered audio paradigm, a concept as seemingly simple as basing a score's tempo on game data becomes rather difficult and unwieldy in practice. However, if the score is procedural, the change could be made dynamically and instantaneously, simple in execution and requiring no extra or redundant assets.

When it comes to timbre and sound design, pre-rendered audio is also very limiting in creating truly adaptive scores. Say, for example, that a composer wants the timbre of a musical sound effect to vary based on how hard the player collides with a certain game object, such that the harder the collision, the louder and brighter the sound should be. To accomplish this using techniques available for manipulation of pre-rendered audio, the composer would have to use a combination of multiple bounces of the sound effect sound file and real time application of audio effects like gain and EQ.[7] Because the sound can only be that which was rendered beforehand, it can vary in timbre only so much depending on the strength with which the object is hit. If the sound is being generated procedurally, on the other hand, the composer/sound designer could create an algorithm that can accept any amount of detailed data (e.g., the velocity of the swing, the player character's current strength level, the size and consistency of the object being swung, the material of the object being swung, the material against which the object is swung, atmospheric conditions, etc.) and process these data immediately to create a sound effect that perfectly takes into account all of the relevant parameters.

Looking at these use-cases, proceduralism is clearly the better technique for any game score that aims for accuracy. However, procedural audio is still a new and largely untapped field in game audio and music design; it is very CPU-intensive and is thus often not feasible for use in most games, since a great deal of the available computational power is reserved for other tasks. While procedural audio may be used to create dynamic ambiences (for example, weather sounds), current gaming technology (as of late 2019) is still not advanced enough to be able to easily procedurally implement any sound a game could possibly require with perfect accuracy using real time methods, while also avoiding the potential to overload the game hardware. That said, game audio industry engineers and designers are creating tools that make the current implementation of procedural audio elements increasingly feasible, effective, and even easy. As this trend continues, novel music scenarios made possible by the paradigm of procedural music are appearing increasingly more regularly.

The Origins of Procedural Music

As we consider the origins of procedural music, it is important to note that the algorithmic techniques used in games today originated far before electronic computers even existed. As scholar Elizabeth Medina-Grey points out in her dissertation "Modular Structure and Function in Early 21st-Century Video Game Music,"[8] many such ideas originated with eighteenth-century musical dice games, wherein rule sets were created for players to create algorithmic compositions. Medina-Grey describes how the creator of the game would compose a series of predetermined musical segments, or *modules,* as well as a rule set that would ultimately guide—based on the rolls of the dice—the musical outcome. In fact, most music scholars believe that even Mozart experimented with algorithmic music techniques at some point in his short, though prodigious, career. In his 2011 article "Algorithmic Composition: Computational Thinking in Music," published in *Communications of the ACM* magazine, Michael Edwards points out that Mozart's "Musikalisches Würfelspiel"[9] uses musical fragments that are to be combined according to the roll of the dice.[10] Music with facets governed either literally by the roll of a die or metaphorically by some other similar random selection algorithm is called *aleatoric*[11] music.

In the following musical eras, algorithmic music techniques continued to develop as composers continued to experiment. Arnold Schoenberg developed and codified musical *serialism*, a method of composition based on cycling through a series of musical elements rather than more traditional considerations of functionality and form. Schoenberg himself codified *dodecaphonic* (or *12-tone*) *serialism* via the use of *tone rows*—a technique by which the composer places each of the 12 tones of the Western chromatic scale into some arbitrary linear order. By the 1940s, composers had developed *total serialism*, in which the serialist techniques pioneered by Schoenberg were applied also to timbres, note durations, intensities, dynamics, and other elements. Randomization was being applied to increasingly granular elements of many musical compositions; composers like Milton Babbitt in the United States and Pierre Boulez in France contributed particularly seminal works to the movement.[12]

Mid-twentieth-century composers like John Cage, Terry Riley, and Brian Eno experimented with giving both musicians and audience members power over their compositions, giving generative music a strong element of human-driven randomization. Terry Riley's composition "In C" (1964)[13] consists of a collection of very short musical loops and a rule set that designated how and when those loops could be played. This performative system allows—even forces—any ensemble performing "In C" to generate a new version of the piece each time. ▶(10.1)

In many cases, including those of "Musikalisches Würfelspiel" and "In C," the aleatoric experimentation consisted of the generative arrangements of the *musical modules*, short musical segments prewritten by the composer, by realization by the musicians at the time of performance. Medina-Grey examined this concept of musical modularity as compositional technique in her thesis, and she concluded that this is arguably the core concept by which early generative music was created. She points out that to this day, it remains a powerful tool when composing algorithmic music for video games: "[in] video games, the modules are often distinct containers of musical data—digital files—and the rules are programmed into the game in the form of triggers and other *if-then* conditions."[14]

The key difference between the generative concert music and the generative game music, of course, is that the aleatoric aspects of game music are produced by computerized randomization rather than human randomization. Procedural game music, then, is realized by an algorithmic combination of some game data, rather than by literal rolls of a die. This causes (unfounded) concern in some composers, as Michael Edwards points out: "Much of the resistance to algorithmic composition that persists to this day stems from the misguided bias that the computer, not the composer, composes the music."[15] In fact, an algorithmic computer program might arguably produce a more personal and subjective resultant composition than a human-based randomization process would, *because* the composer would have a great deal of control over how the computer processes the relevant information. It is a human who ultimately creates the rules by which the data-based computer algorithms abide. Thus, game music composers still have a deep level of control over how their algorithmic music sounds. Just as Eno and Riley composed systems for humans to follow, game music composers create rules for how computers use game data to generate music.

One concept integral to generative composition is that of *randomness*. As the roll of a die introduces randomization naturally, algorithms too insert randomness into procedural game music. Algorithmic composition therefore requires the composer to design the "grammar" of the algorithm. The computer, given rules to follow, builds a particular sonic vocabulary and performance aesthetic.[16] In some instances, computers can do this intelligently using a computational paradigm called *machine learning*.

Designing Procedural Music Systems

Machine Learning

M*achine learning* is a budding field that focuses on the teaching of computer systems to take information inputs and "learn" by recognizing patterns in the collected data—much like the animal mind takes stimuli as inputs, categorizes those stimuli as experiences, and is then able to make decisions based on that learned information or pattern.

Machine learning allows some conceptually simple but potentially highly useful applications for audio implementation in video games. For example, a computer can very easily determine if a certain section of audio is "silence" simply by checking if the audio signal has a value of "zero." By measuring the *zero crossing ratio*, the number of times that an audio signal crosses the zero-value in a given amount of time, a computer can then determine whether there is a percussion element in a given piece of audio (for if there were, the wave would cross zero much more frequently than if there were no percussion). In fact, this same capability of machine learning allows the beat-and-tempo detection function common in many modern DAWs.[17]

In another common use of machine learning, the computer observes data, determines potential patterns, and decides how to mimic or reproduce those patterns of data. Often, this is accomplished by utilizing *Markov chains*, a model of sequencing probable events based on the states of previous events. Though Markov chains are now a largely obsolete machine learning process, they still prove the most useful model for "learning" short melodic phrases. Modern machine learning is generally rooted in neural networks; essentially, complex multilayered algorithms that recognize patterns. These algorithms are designed based upon how neuron cells in the animal brain function and cooperate, making neural networks an important part of the growing field of AI.[18]

Understanding these machine learning techniques helps us see how they are related to the way in which a machine might analyze music. For example, if a computer were "fed" a series of melodies written by a single composer, it could then analyze elements such as how often there is rhythmic deviation from the quantized grid, intervallic structure, key center, chromaticism, and so on; based on the results of the analysis, it then could determine the stylistic aesthetics of that composer's work. Finally, it could take those stylistic aesthetics and produce music in that style. In other words, though this music would not be literally written by the composer, the composer would be the primary influence over the generated music.[19]

While until recently, creating a procedural score has required knowledge of programming, some audio programmers and engineers are using the principles of machine learning to create software that allows composers to write procedural music without

Advanced Reactive Music Concepts

having to necessarily also have a deep-level understanding of visual or textual programming languages. Instead of requiring the implementation of a complex bespoke algorithmic music system, this kind of software uses machine learning to analyze and reproduce a composer's style.

One company developing such generative software is engineer Daniel Brown's Intelligent Music Systems (IMS),[20] which specializes in creating software that allows composers with limited programming knowledge to create generative music for games. Their software, the Dynamic Percussion System (DPS), is a procedural percussion engine featured prominently in the score to Crystal Dynamics's *Rise of the Tomb Raider* (2015).[21] The DPS engine analyzed examples of the compositional style of the game's composer, Bobby Tahouri, and creates music in real time during gameplay based upon the results of that analysis.

Similarly to how middlewares like *Wwise* and *FMOD* facilitate audio implementation in game engines,[22] the IMS software offers a visually appealing and intuitive interface through which composers can design procedural music. The UI, seen in Figure 10.1, resembles a basic mixer window in a DAW. First, the composer feeds the DPS example pieces of music to analyze and emulate. Then, after the composer has added the music examples, the DPS analyzes it, learning the patterns, structures, textures, and overall style

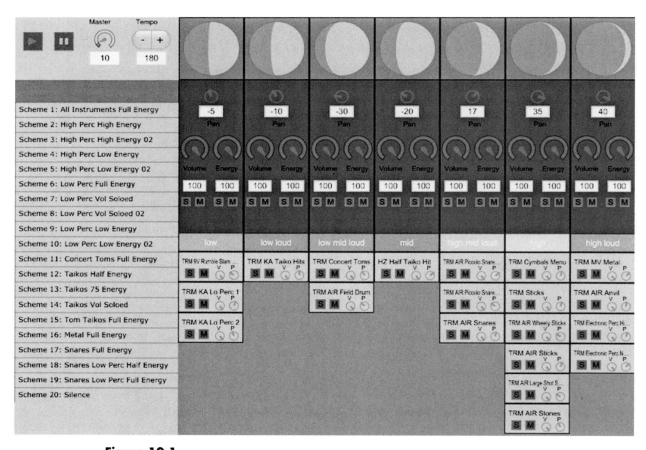

Figure 10.1.
DPS UI design by Daniel Brown and Camellia Boutros.
Image has been provided with permission and is a screen capture of the *DPS* taken by Daniel Brown.

and aesthetic of those pieces. All of the data collected from this information are then saved into a single large file. The DPS mixer then uses this data file to realize improvisational content directly based on the composer's original material. Finally, at the end of the process, the composer has the ability to further refine the computer's stylistic aesthetics and subsequent compositions through the UI, choosing which musical elements and moments sound "best." The DPS will then integrate the chosen music into the game, realizing elements in direct response to the events of gameplay. As Brown states, "It's like giving the music to an ensemble, then letting them jam out on it in the game instead of playing it as written."[23] Thus, though the music is procedural, the composer exerts creative control over what the DPS produces, from the writing of the initial music to the implementation of the final elements.

Indeed, the DPS offers a simple UI for procedural integration of percussion in games. However, the DPS engine is limited to percussive elements. In order to also create melody and harmony algorithmically, composers must turn to other methods—at least until Brown or one of his cutting-edge peers invents an algorithmic engine that can handle melody and harmony. Until that time, composers are best served by understanding the fundamentals of visual programming.

Visual Programming Languages

A well-designed procedural system provides composers with great compositional freedom. They can effectively use any game data they desire to affect the music in any way that they want. However, at the technical level, procedural systems are still fairly difficult to create and implement in a game engine. Thus, currently, the use of a visual programming language (such as Max MSP or Pure Data) offers composers a relatively straightforward solution. Visual programming languages allow composers to piece together visual representations of specific functions. The syntax (or grammar) of the programming language itself is composed of *objects*, represented graphically in the UI as small modular boxes, as opposed to text like many other programming languages. Each object has a specific purpose; some represent *functions* (which carry out a set of instructions, or, in other words, apply an algorithm) and others are *variables* (which specify the value of a piece of information). By combining certain functions with certain variables in certain ways, composers can use these objects to "write" a program. In the case of a visual programming language, "writing" means to connect a series of objects into a logical diagram to carry out a specific process.

Understanding how to connect objects in a visual programming language is a very similar concept to that of understanding analog signal flow in a recording studio or MIDI signal flow in *Logic Pro X*'s Environment; each object receives information via an input and/or sends information through an output. Thus, for example, information from Object A's output can be routed to Object B's input. In the case of Max MSP, these inputs and outputs are called *inlets* and *outlets,* while the "cables" that connect them are called *patch cords*.

Visual programming languages are also generally semiotically more intuitive than textual programming languages. For example, instead of a `PlaySound()` function being written in a script, a visual programming language might instead have an object

Advanced Reactive Music Concepts

Figure 10.2.
ezdac~ object in MaxMSP. Zicarelli, David. "Max MSP." 8.1.2, Cycling '74, 0AD.

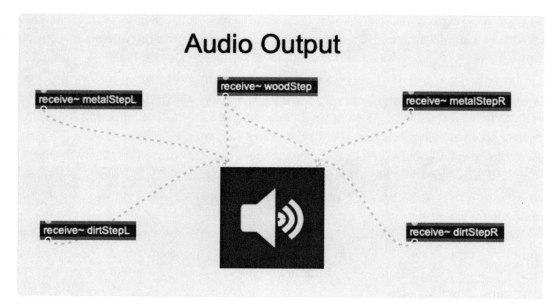

that visually resembles a speaker. This speaker will have an input and will receive a signal from another object.

Take the following example coded in Max MSP (Figure 10.2): The object that looks like a speaker is called an "ezdac~" and is used for outputting audio through the computer's sound card. This ezdac~ object is receiving different types of footstep sounds into its inlet. Whichever footstep is sounding at any given time will, outlet to patch cord to inlet, play through the ezdac~ object, which is directly routed through to the audio output.

Visual programming languages are useful for designing algorithmic systems because they allow composers to create visual representations of their audio systems, which is a paradigm already familiar to composers familiar with DAWs and plugins. Thus, for composers, using a visual programming language is often preferable to designing an algorithmic system in a code-based textual programming language, which could be more difficult to navigate and implement intuitively. Because of the familiarity most game music composers already have with the visual logic paradigm, it follows that the more the UI of a visual programming language resembles that of a DAW, the more accessible creating algorithmic music for games would become to them.

Procedural systems generally operate using MIDI. This creates both advantages and disadvantages when it comes to implementation. The main advantage, of course, is that MIDI can provide the music system with a great deal of information on how to dynamically change the music in real time based on game data. For scores with an electronic music aesthetic, this is a wonderfully liberating method of implementing music in a game. However, this ability becomes a double-edged sword when the score requires the use of complex sampled acoustic instruments. Using MIDI to perform complex acoustic sample libraries uses a large amount of a computer's real time processing capabilities. Audio budgets (how much hardware power is allotted to the audio system of a game) for games already tend to be somewhat limited, so introducing an orchestral library of samples being played by a procedural engine in real time is simply not feasible in most cases.

On the bright side, this limitation may not last much longer. Professionals like Daniel Brown continue to invent new ways to improve the implementation process of fully procedural scores. Some companies, like Elias Software, are working on game-optimized sample libraries and engines, making the use of orchestral samples far more feasible for procedural audio. In the not-so-distant future, should trends continue, it is likely that increased hardware power and improved software solutions will usher in an age of fully procedural, fully orchestral/"acoustic" game music scores. With this in mind, understanding the possibilities and implementations of procedural music will become increasingly important for composers of music for games. In Chapter 11, we will explore the many musical possibilities that an algorithmic audio system can provide in further detail.

Procedural Music Systems in Action: *Rise of the Tomb Raider*

Let us now observe some practical applications of procedural music systems by taking a closer look at the development of *Rise of the Tomb Raider*. According to Phil Lamperski, the lead audio designer who spearheaded the implementation of how the DPS would function in the game, the development team determined a four-state combat music system early in the development process:[24]

1. Whenever the player, as game protagonist Lara Croft, is sneaking around and undetected by enemies, she is in the Stealth state. Lamperski programmed the music for this state to be ambient, favoring sparse, low-pitched percussive elements. ▶(10.2–10.3)
2. The Low Alert state occurs when enemy NPCs[25] become aware of a possible disturbance, but do not yet realize the threat. In this state, the music rises in tension slightly by using more high percussion and tense musical elements than the first state, creating the anticipation of possible detection. ▶(10.2–10.3)
3. The High Alert state builds in intensity even further—at this point, the enemies are now fully aware of a threat and are actively searching for Lara Croft. In this state, the high percussion continues, introducing medium-intensity low drums to put the player further on edge. ▶(10.4–10.5)
4. Finally, the fourth state, when Lara Croft is finally discovered by the enemies, is full-out Combat. In this state, the music features huge, low-pitched, high-impact rhythmic drum beats—a jarring musical effect, portraying the high-stakes environment of a life-or-death situation. ▶(10.4–10.5)

Within these four main combat states, Lamperski also created a series of "sub-states." Some examples of sub-states that have an impact on the realization of the music are Concealment (the player hides Croft in a bush), Finisher (the player executes a *coup de grâce* on an enemy NPC), and Equips Bow (the player pulls out Croft's bow and arrow). This intricate state system is complemented by Brown's intricate procedural music system, which has the ability to dynamically interpolate states in real time.

Most systems use audio crossfades to smoothen musical transitions, but the DPS uses the incoming game data to instantaneously react and adapt—for example, instead of crossfading to a fixed cue of higher intensity, DPS would merely build dynamically and musically. Though this effect is algorithmically generated, this transitioning of DPS is still closer to the affect of a human musician reacting in real time to the on-screen events. As Lamperski describes this capability:

> The interpolation feature of the engine is very powerful, especially when coupled with its immediate transition functionality. For instance, there would be cases when [Croft] would immediately transition into a percussion scheme that would [be interpolating] between two schemes. So when [Croft] starts in the stealth percussion scheme it is always interpolating between low energy quiet drums to higher pitched higher energy drums as you get closer to an enemy, but it can change immediately if the player executes a finisher, or goes into the bushes, etc. Once the player comes out of the bushes or the finisher state, the percussion will immediately transition back into interpolating between two schemes (assuming the player goes back into stealth). All of these changes are very subtle, but work great [sic] to score the narrative of the player.[26]

Unlike the procedural expanses of *No Man's Sky*, which we discussed earlier in this chapter, the generative elements of *Rise of the Tomb Raider* serve a game with a predetermined narrative arc. The story moves linearly (as in the linearity of the plot points, or series of events, and not as in the *non*linearity of interactivity afforded by the actions and processes of gameplay), each level being designed specifically by the developers to function in a particular way. While creating a score for the procedural game system design of *No Man's Sky* would have been rather difficult to effectively realize without the use of a generative audio system, the state-driven gameplay of *Rise of the Tomb Raider* could have been realized by any number of scoring and implementation techniques. However, the potential reactive and interactive power of a well-implemented procedural music system was too great for the Crystal Dynamics team to pass up; by using DPS, they enabled many reactive and adaptive score possibilities.

Again, despite the linearity of the *Rise of the Tomb Raider* storyline, the gameplay itself is still event-based and interactive, and therefore imbued with some degree of unpredictability. For example, it is not possible for Tahouri to have predicted precisely when the player will come across an enemy, how or when the player will have engaged that enemy, or even whether the full Combat state would be triggered. He could not know at which moment the player might raise Croft's bow, if an arrow might be released, or if that arrow might hit or might miss an enemy. Each millisecond, each possibility leads to a different result (and, therefore, a different score state). By utilizing the DPS, Tahouri was able to provide *Rise of the Tomb Raider* with a highly reactive score solution, well crafted to each individual play-through experience.

Conclusion

In music written for video games, a procedural score is created by implementing an algorithmic music system that responds to game data and generates audio in real time based upon the rules initially established by the composer. Because of the large computational load required to "realistically" use sample libraries of acoustic instruments, using such a system is most effective when the score's aesthetic is primarily electronic. In the near future, however, computational power and storage space may increase enough to the point that fully realized "acoustic" procedural scores become more feasible and, thus, more common.

Creating a procedural music system generally requires knowledge about how to use a visual programming language to create an algorithmic interface for establishing rulesets. However, software like DPS uses machine learning to offer composers a simple user interface for integrating procedural percussion into games, and, again looking into the near future, engines that likewise convincingly handle melodic and harmonic elements may be within the reach of tomorrow's composers.

Exercises

1. Choose a game that does *not* use a procedural audio system. Analyze its soundscape, determining how the audio is integrated. What are the key features of the music design? To what degree is the music interactive? What are some of the strengths and weakness of the design?
2. Using this same example, create a hypothetical procedural music design for the game. What advantages would using a procedural system offer this game's music? What disadvantages would come along with the generative approach?
3. If you were to be tasked with re-creating this game's soundscape, would you use a procedural score, or not? Explain your answer.

Notes

1. GameSoundCon.
2. Thomas Dolby, "The Next Generation of Non-linear VR Composers," GameSoundCon 2018 (convention lecture), Millennium Biltmore Hotel, Los Angeles, CA, October 10, 2018.
3. Game Developers Conference.
4. Real time synthesis may be used, for example, to create modeled instruments, functioning particularly well for modeling string and wind instruments. Also, many older game hardware systems, which we discussed in Chapter 3, generated audio using real time synthesis, often on a specialized audio chip (like the TI SN76477 of *Space Invaders* or the MOS Technology 6581 Sound Interface Device "SID" chip of the Commodore 64). However, in most of these cases, these musical scores were fully pre-composed, the music acting as underscore despite being generated in real time.
5. Paul Weir, "The Sound of 'No Man's Sky,'" Game Developers Conference 2017 (convention lecture), Moscone Center, San Francisco, CA, March 3, 2017.
6. That is, in non-real time.
7. Equalization.
8. Elizabeth Medina-Grey, "Modular Structure and Function in Early 21st-Century Video Game Music" (PhD diss., Yale University, 2014), 18–20.
9. In English, "Musical Dice."
10. Michael Edwards, "Algorithmic Composition: Computational Thinking in Music," *Communications of the ACM*, July 2011, cacm.acm.org/magazines/2011/7/109891-algorithmic-composition/fulltext.
11. From the Latin *alea*, "die." As Julius Caesar said as he crossed the Rubicon River in northern Italy on his way back to Rome from a successful military campaign in Gaul, without having first disbanded his troops: "*alea iacta est*" ("the die is cast")—in essence, "what's done is done." Once the river would be crossed with a standing army, a civil war (albeit one which would

culminate in his ascension to the position of Dictator for Life, and ultimately the establishment of the Roman Empire) would become inevitable.

12. Barbara Russano Hanning et al., *Concise History of Western Music* (New York: W.W. Norton, 2009), 627.
13. Terry Riley, *In C* (New York: Associated Music Publishers, 1964).
14. Medina-Grey, "Modular Structure and Function in Early 21st-Century Video Game Music," 19.
15. Edwards, "Algorithmic Composition: Computational Thinking in Music."
16. The process by which data are received from an input and then output as sound or music using an algorithmic software system is called *data sonification*. This process is, in many ways, what is occurring when the data from games are sent to an algorithmic music system, and then converted into a composition.
17. John Byrd, "Introduction to Machine Learning for Game Audio," GameSoundCon 2018 (convention lecture), Millennium Biltmore Hotel, Los Angeles, CA, October 10, 2018.
18. Artificial intelligence.
19. This particular use-case is starting to become rather visible because of the question it raises: if a computer is used as a compositional tool, and is trained on the music of a particular composer as such, who owns the music that computer subsequently produces? Is it the composer whose style was analyzed? Is it the developer of the machine-learning algorithms? Is it the computer itself?
20. "Intelligent Music Systems," Intelligent Music Systems, accessed November 4, 2019, http://www.intelligentmusicsystems.com
21. *Rise of the Tomb Raider*, game disc, developed by Crystal Dynamics (Tokyo: Square Enix, 2015).
22. The beauty of Brown's DPS is that, much like middleware bridges the gap between game engines and audio implementation, his software facilitates the inclusion and implementation of procedural music, providing a user-friendly UI that composers can use without having to understand the "back end" of the programming. In other words, it allows composers to focus more on musical matters than technical matters.
23. Daniel Brown (designer of DPS), interview with author, November 12, 2018.
24. Phil Lamperski (lead audio designer of *Rise of the Tomb Raider*), interview with author, December 6, 2018.
25. Non-player characters.
26. Lamperski, interview with author, 2018.

11

Advanced Algorithmic Music Systems

Introduction

Without question, the many tools we have discussed so far provide composers and developers with excellent methods of creating advanced, dynamic musical scores. In Chapter 9, we saw how the combination of horizontal resequencing and vertical layering can be used to create a detailed, modular system that can account for a vast array of gameplay possibilities. However, even a complex system like the music matrix of *Just Cause 4* (2018)[1] is limited to pre-rendered audio segments and loops. Building on our discussion of procedural music, in this chapter we will learn about the advanced capabilities of custom-built algorithmic music systems, and how they can respond to minute changes in game data. For this exploration, we will use *Uurnog Uurnlimited* (2017),[2] a game developed by Niklas Nygren (aka Nifflas), which features such an algorithmic music system. Nifflas designed *Uurnog* with real time audio generation as a key feature of the gameplay experience, even going so far as to create a stand-alone music system called *Ondskan* to give himself nearly unlimited control over how a procedural musical score is attached to game data.

Through our exploration of advanced reactive music systems, we will encounter many tools useful in thinking outside the box when crafting a music design for a game. These tools can give developers multiple approaches for creating scores that adapt to each play-through and provide the player with a uniquely personalized score.

Creating Advanced Reactivity with an Algorithmic Music System

Remember, algorithmic music is procedural and is generated in real time, meaning that practically any element can be changed immediately without constraints. While our chapter on procedural music introduced us to some interesting generative scoring techniques, demonstrating how easily we can create procedural music thanks to interfaces like Daniel Brown's Dynamic Percussion System, we will be diving further into algorithmic composition in this chapter, exploring how a custom algorithmic music system can open up even more musical possibilities. *Uurnog Uurnlimited* is a prime example of an indie game that pushes the boundaries of what composers can accomplish with a custom procedural music engine. Nifflas's work takes advantage of the plethora of parameters that can be used to manipulate algorithmic music, turning them into creative rule sets to determine just how the music will react to player data. In many cases, advanced reactive music systems like these imprint the composer's personality into the gameplay. We will explore *Uurnog*'s music system, *Ondskan,* in great detail later in this chapter; first, let us dive into the possibilities offered by algorithmic systems in general.

Compositional Devices: How to Use Them with Game Data

The field of algorithmic music is a broad compositional sandbox. Rather than making absolute decisions in the compositional process, the composer gives the computer musical agency, thereby essentially collaborating with the system. For the music system in *Just Cause 4*, Zach Abramson composed and pre-rendered discrete "building blocks" of music, which were then implemented modularly. Based upon game data from player actions, the *Just Cause 4* music system assembles these modules to form the score; many music systems in modern games operate similarly.[3] However, using an algorithmic music system allows the compositional possibilities to go further, not only piecing together extant musical modules but actually manipulating and changing them in real time.

A key tenet of algorithmic game music is the idea that any game input data can be used to affect almost any musical element. The real time capabilities of an advanced reactive music system allows music to be manipulated with great detail. As we saw in Chapter 10, this grants the opportunity to manipulate elements like form, pitch, timbre, rhythm, harmony, and so on. In this section, we will therefore explore a detailed, albeit incomplete, list of musical devices that could potentially be attached to game parameters. We will also see some ideas on how these devices might be attached to game parameters.

Compositional Devices for Manipulating Notes in Real Time

At the most basic level, a *note* is a musical element with discrete pitch and duration. Notes can be manipulated in many different ways; throughout the history of Western music, a variety of compositional devices have been used for this purpose, such as contrapuntal devices, which provide excellent ways of building interaction between various themes and melodies in a composition. Recreating these compositional devices in software gives composers a way to use established compositional techniques to build the music in response to the on-screen action.

Ultimately, these musical devices must be connected to game parameters to enable real time musical reactivity. The following is a list of common contrapuntal and compositional devices used to manipulate melodies, as well as some examples of how they might be connected to game parameters. The example we'll use to illustrate these concepts is the "Main Theme" from *Grobo*, by PHÖZ. ▶(11.1)

- *Repetition*: The reiteration of a sequence of notes (e.g., a melody or a theme) (Figure 11.1). This can be quite useful in the context of a reactive music system for instances in which the composer wishes to create an association between a musical event and a player action. For example, a melody could be programmed to repeat whenever a specific action is carried out, strengthening the player's mental and emotional associations with the performance of that action. As in concert music compositional practice, continued repetitions may require slight modifications to avoid monotony; such a modification is trivial for advanced music systems to enact.

- *Sequence*: Transposing a melody or phrase while maintaining the same intervallic structure (Figure 11.2). A game composer might designate a certain in-game event to trigger sequential motion in the music score; for example, if the player triggers a mini-game in which he has exactly 10 seconds in which to ascend a challenging platform area and recover an item, this trigger could also be used to cause the music to "grab" the last four notes directly before the trigger and ascend sequentially until the mini-game is over.
- *Retrograde*: Performing a musical line backward from end to beginning. In an advanced system, retrograde can be used in a multitude of ways, like to signal the player's progress in a level; if the player were to lose ground, the melody might sound in retrograde (Figure 11.3). Generally speaking, when both the melody and rhythm are in retrograde, this is referred to as *total retrograde* (as per Figure 11.3). However, one can also use only *melodic retrograde* (the melody is in retrograde, but the rhythm remains the same), or *rhythmic retrograde* (the rhythm is in retrograde, but the melody remains the same).
- *Inversion*: Performing a musical line upside-down—that is, whenever the original interval between notes moves upward, the inverted version will move downward, and vice versa (Figure 11.4).
- *Retrograde inversion*: Performing a musical line upside-down and backward (Figure 11.5).

Figure 11.1.
The "Main Theme" from *Grobo* notated in its original form.
Composed by PHÖZ and all musical devices notated by Alba S. Torremocha for use with permission in this text.

Figure 11.2.
The "Main Theme" from *Grobo* sequenced.

Figure 11.3.
The "Main Theme" from *Grobo* in retrograde.

Figure 11.4.
The "Main Theme" from *Grobo* inverted.

Advanced Reactive Music Concepts

- *Diminution*: Halving the rhythmic value of the notes of the melody (Figure 11.6).
- *Augmentation*: Doubling the rhythmic value of the notes of the melody (Figure 11.7).
- *Variation*: The elements making up the original melody are modified or introduced in a different manner (Figure 11.8). Variation is a very common and useful compositional technique. Though in this section, we are referring only to the variation of a melody, variation of harmony or musical form as a whole is also an exceedingly useful tool for composing video game music.
- *Contrast*: Creating a musical statement that "answers" the "question" asked in the preceding phrase—the "consequent" of an "antecedent/consequent" relationship (Figure 11.9). For example, in jazz, the 12-bar blues form is based on this relationship: the first four bars pose a question, the second four elaborate on this question (often via *variation*), and the last four answer it. Therefore, in the 12-bar blues, the last four bars would be the contrast.

Figure 11.5.
The "Main Theme" from *Grobo* after retrograde inversion.

Figure 11.6.
The "Main Theme" from *Grobo* after diminution.

Figure 11.7.
The "Main Theme" from *Grobo* after augmentation.

Figure 11.8.
A variation of the "Main Theme" from *Grobo*.

Figure 11.9.
A contrast of the "Main Theme" from *Grobo*.

Compositional Devices for Manipulating Harmony *in Real Time*

Harmony is another underlying element of most music influenced by Western musical traditions. In *Just Cause 4,* the majority of the game's music is written either in the key of D, or a key closely related to it. This ensures that the music system can transition between different keys easily without any jarring musical changes. This exemplifies a major musical limitation of pre-rendered audio: in order to keep transitions smooth, the music becomes limited to specific musical keys. In an algorithmic system, however, the music can exist in any key at any time and can transition easily between various states, all because the notes are not "baked" into a file. Reading through the following list of harmonic tools, consider how each of these devices could be attached to a game parameter and become a reactive part of a game's musical score.

- *Chordal modifications*
  - *Changing notes*: Notes of a chord are reordered or replaced, changing the overall sound-color or even musical function of that chord. In a game where the player chooses dialogue options, the composer might set certain notes in a chord to change depending on the implications of the chosen answer.
  - *Adding notes*: Notes are added to a chord, giving it a new character. This can be done either in the core section of the chord, or in the form of extensions.[4]
  - *Inversion*: A note other than the root of the chord is the lowest note in the chord.
  - *Arpeggiation*: Performing the notes of a chord consecutively instead of simultaneously, as is often found in harp music (hence the name). An especially common sound in game music because early systems were only capable of monophonic sound production, so chords were arpeggiated in order to outline the harmonies.
  - *Rotating the notes of a chord*: Changing the order of the tones, but keeping the notes the same. Of course, changing the order of notes in a chord has the potential to change the character of that chord quite a bit, but it also keeps the structural integrity generally functional within the key.

- *Progressional modifications*
  - *Chord substitution*: Replacing one chord in a progression with another chord, thereby changing the bass note of that chord in the progression as well. A chord substitution may be appropriate as the next chord in a progression with a darker option if, say, the player should take damage.
  - *Reharmonization*: Replacing one chord in a progression with another chord while maintaining the bass note of that chord in the progression. In essence, this changes the function of that bass note in the context of the harmonic progression, even though that note itself has not changed.
  - *Adding chords*: Creating a denser musical arrangement by adding in more chords to the progression. This often also necessitates a faster *harmonic tempo*.[5]
  - *Subtracting chords*: Creating a sparser musical arrangement by removing chords from the progression. As this often also necessitates a slower harmonic tempo, the remaining chords become more musically prominent because more time is spent on them.

Compositional Devices for Manipulating Rhythm in Real Time

In many cases, the underlying rhythm of a composition may be the central focus of that piece, defining its character and intensity, and suggesting a certain emotion. As we saw previously, *Just Cause 4* uses different rhythmic densities depending on the intensity of the on-screen action. The following list of ideas barely scratches the surface of the possibilities inherent in procedural rhythmic manipulation:

- *Tempo*: Changing the tempo of a piece of music, drastically altering how it is perceived. Even a minute change in the bpm is noticeable; larger changes can be the difference between a track being relatively relaxing or stressful (tempo was, of course, the very parameter attached to enemy proximity in our earlier *Space Invaders* example).
- *Adding beats*: Simply, adding beats to the musical idea. One use in a game focused around collecting a certain type of object could be adding a beat to a bar of music each time one of those objects is collected.
- *Subtracting beats*: The opposite of adding beats. One might subtract not only the chord (as in the *subtracting chords* point noted earlier), but even the beats in the bar where that chord existed. This could be an interesting way to account for, say, the player running out of moving space as an enemy grows closer; each time the enemy approaches by a certain amount of distance, the bar drops a beat, until finally the music runs out, vanishing into nothing, when the player runs out of space and the enemy wins.
- *Changing time signatures*: Adding and subtracting beats generally necessitates this action, but time signatures can be changed for other reasons, too. Modifying how many note durations exist in each bar greatly affects the rhythmic feel of the music. As a hypothetical procedural example, the time signature of a cue could depend on which character you are currently playing as. If that character's personality is more awkward, one could use a more "odd" time signature, whereas if the character is confident, a strong time signature like 4/4 could be used.
- *Swing*: Broadly, the difference in durational quality between the beats and the up-beats. In many DAWs, the amount of "swing"[6] feel can be easily modified, giving the music a different rhythmic character. As an example: in many RPGs, the player has the option to drink alcoholic beverages. Turning up the swing-amount of the rhythm as that character becomes more drunk could mirror the character's inebriation with the rhythmic "looseness."
- *Density*: What degree of detail or busy-ness is utilized on the rhythmic grid. Is the rhythmic pattern devised of quarter notes, eighth notes, sixteenth notes, or another rhythmic pattern? The smaller the value, the denser the rhythm will be.

Compositional Devices for Manipulating Timbre in Real Time

Timbre is a defining property of any musical instrument. Making timbral changes in real time is an interesting way of changing an instrument's purpose in the music. Each synthesizer, sampler, and MIDI environment provides composers with numerous ways of affecting timbre within the context of instrument design. Synthesizers even provide different initial waveforms. A variety of effects may also be used to change the sound of an instrument.

- *Manipulating timbre*: One can use filters or simply modulate parameters on the current synth or sampler to change the timbre of an instrument. Altering an instrument's timbre might make it sound like a different instrument altogether. For example, take an oboe that is playing a character-theme. One way of implementing a timbral manipulation might be to attach a low-pass filter to the oboe, filtering out its upper spectral content and leaving it sounding much rounder and less piercing. If the character is relaxed, his theme remains filtered, but his anger grows as the player makes mistakes. The more his anger grows, the more high frequencies the low-pass filter allows through, directly connecting the character's anger to the oboe's brashness.
- *Instrument*: Cycling between different instruments altogether, replacing one timbre for another when certain in-game events occur. For example, there might be a locational melody on-screen that always plays in that particular location. When the player encounters a different character in that location, the instrument might switch to reflect that change, giving each character a unique sonic signature despite not having original themes.
- *Velocity*: The classic MIDI parameter, describing the initial dynamics of the resulting notes. Depending on how hard or fast the MIDI instrument is played, the note will sound louder and brighter when played harder, and quicker or softer and rounder when played more softly.
- *Audio effects*: Numerous audio effects that music producers have used for decades to manipulate sounds on records can be used to manipulate the timbre of a sound in a game. To name but a few: distortion, ring modulation, bit reduction (commonly used in modern game music to achieve the classic low-bit game music sound), delay, and reverb.

Compositional Devices for Manipulating Form and Structure *in Real Time*

The structure of a piece is one of the most commonly altered properties in the context of functional game music. As we saw with *Just Cause 4,* modular music systems make structure both variable and gameplay-dependent. Just as with other musical elements, however, using an algorithmic music system provides an added layer of composer control with which to dynamically affect the form of a piece. A familiarity with traditional composition form and structure techniques better informs the construction of a music system that can react dynamically to player input while also retaining the structural integrity of the music. While there are numerous formal possibilities, here are two basic, overarching formal techniques:

- *Variation*: Using many or all of the aforementioned devices together, the composer gains the compositional power to recreate musical elements in entirely different contexts. In other words, the player can create a *variation* of a musical idea. This is an incredibly powerful and personal compositional device and can be used in many different ways, with as many styles of variation as there are composers.
- *Repetition*: Here, repeating a section of music, such as a theme, as needed. A simple example might be found in the context of a battle. The enemy's theme could be repeated over and over with slight variations each time, becoming more and more hectic as health runs out. Finally, once defeated, the enemy theme might sound "broken down," then finish, the soundtrack continuing on to a new section of the music.

Ultimately, this list of compositional devices is but a glimpse of the multitude of ways composers have allowed real time manipulations of their compositions in music for games. Algorithmic systems provide a wealth of musical, compositional, and creative power. That power, however, is contingent on thoughtful and intelligent implementation strategies.

Implementing an Advanced Reactive Algorithmic Music System: *Ondskan*

As we have seen, algorithmic (or procedural) music systems allow incredible control when it comes to reactivity. Since all the elements of the music are being generated in real time, a well-designed music system can output practically any result based upon game data. This means that player choice can be set up to affect practically any musical element, according to the initial rules established by the composer.

Nifflas did not hold back when it came to creating a custom music system for *Uurnog Uurnlimited*. The game is designed such that even the most minute player action can have a profound effect on the music. An action as simple as the swing of a weapon, for example, might change the key of the music, change the melody completely, or inform any other number of musical curiosities. In one example found in the game, the music changes key each time a bomb explodes. In another, the music's tempo is directly attached to the speed of a small cat running away from the player in the game. All in all, Nifflas's custom system can essentially receive any type of input data, and create any musical result.

Categorizing Musical Results

Any algorithmic music system receives data, categorizes it, and then manipulates it based on the rules established by the composer. With this in mind, it is imperative to determine how that data will be categorized when it enters the system. It is often useful to create custom values that can later be manipulated. For *Ondskan*, Nifflas created a number of "values" that identify and categorize notes and aid the system in reacting to the music according to his rules and his personality:

- *Degree*: Notes can be identified by their specific degree in a scale, which corresponds to a certain number. This makes programming specific outcomes fairly simple within the system.
- *Approximation*: Nifflas uses "approximation" to tell the system roughly where a note should fall within a certain scale, giving the computer some influence over the randomization.
- *Awkward*: *Ondskan*'s Awkward system is a wonderful example of creativity in music design. It picks the closest "awkward" note it can get. One important question to ask here is: *what defines how awkward a note actually is?* In this particular case, "awkward" is a subjective value set by Nifflas.
- *Combos*: Approximation *and* Awkward: This is an example of combining two parameters to further define the note possibilities of real time composition. The system tries to

find a note that satisfies both conditions of being "awkward" and is within the approximation zone of the specified area.
- *Other ideas*: Notes could also be categorized based on their consonance, dissonance, intervallic relationship to the last note, groupings of notes, the ornateness of a phrase (how many grace notes, triplets, sixteenths, embellishments, etc., it contains), or rhythmic values. In truth, the possibilities are functionally limitless.

Turning Game Data into Music Using a Custom Algorithmic System

Nifflas's *Ondskan* system is a clear example of the possibilities that custom algorithmic music software can offer the modern composer of game music. We will now examine some of its functionality to better understand how it works and the musical world that it (and software like it), opens up to the composer.

Ondskan is composed of a series of custom-built modules. Note that here "module" refers not to a segment of music, but to a section of software that has a specific function. Each module serves a unique purpose when it comes to the manipulation of music based upon game data.

Ondskan's most basic modules contain familiar musical tools. For example, the synthesizer or sampler modules, seen in Figure 11.10, produce sound in different ways; these are essentially "instrument" modules. The synthesizer gives the composer a way of generating pitch with waveforms, while the sampler module allows the composer to use and manipulate audio files as the basis of a musical instrument from which to play.

Ondskan also contains modules that serve familiar musical effect functions. For example, the harmonizer module creates musical harmonies for a single note. However, like most of *Ondskan*'s clever systems, it provides the composer with functionality beyond the norm. On top of simply copying a single note and creating harmony from it, those other

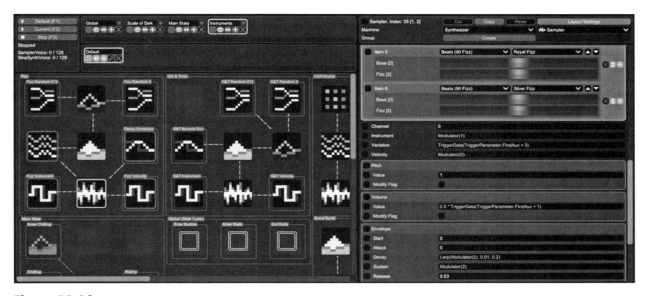

Figure 11.10.
Synthesizer and Sampler modules in *Ondskan*.
Screen captures of *Ondskan* taken by Niklas Nygren and provided for use in this text.

notes can also be shifted in the time domain, essentially turning the notes that would have been stacked upon the first one instead into a linear melody.

Ondskan's Most Liberating Modules

Ondskan even contains a few modules with essential functions that provide compositional liberty far beyond what is attainable with most musical systems:

- *Dry Trigger Module*: This module was created by Nifflas to be a versatile piece of software. It accepts input data, runs that data through its settings, and then generates a trigger message, which is sent directly to a synthesizer telling it what note to play and when to play it; or, perhaps, telling it how to modify a note that is already being played. More specifically, that trigger message contains information about the pitch, scale degree, Awkward (i.e., how "awkward" the note is, according to Nifflas's subjective awkwardness parameters), approximation, and octave values; Nifflas points out that the inclusion of any of these parameters in the trigger message is optional.[7]

  Since the module accepts input data in the form of expressions,[8] the possible uses for this module are various. Nifflas gives us the example that one could set up the module to play a note every sixteenth footstep: to do so, the trigger field would be set using an expression such as `Every(Step(), 16)`.[9] This expression *expresses* the value of every sixteenth step. This module alone offers a valuable glimpse of the possibilities when designing a custom system: its versatility is impressive, allowing the composer to set it to act upon almost any type of input data and then affect notes in the music.

- *Arpeggiator Trigger Module*: While in function similar to *Dry Trigger*, the *Arpeggiator Trigger Module* utilizes an internal counter that increases each time a trigger is sent. In other words, it functions almost like a *for* loop,[10] allowing the composer to loop an action until a maximum value is reached. This maximum value can be set within the module's settings.

- *Scale Module*: This module allows the composer to define custom musical scales. This incredibly useful tool expands the compositional possibilities greatly when it comes to algorithmically focusing on a certain harmonic key center or how to form a melody. Setting a custom scale also allows the manipulation of melodies based specifically on that scale, giving an impressive amount of control.

- *Shuffle Modulator Module*: As Nifflas puts it, this particular module (as seen in Figure 11.11) "takes a sequence and shuffles it."[11] In other words, any sequence of data from any other module can be shuffled and output in a random order by using this module. If the composer wishes to shuffle the order of notes in a scale or melody, they would use this module to do so. Nifflas points out that this module is particularly useful for shuffling around the sequence of events in a drum loop, making it unique each time it comes around.[12]

Even beyond these incredibly powerful tools, *Ondskan* still boasts yet more useful and unique modules. The *Markov Module* is based on the concept of *Markov chains*[13] and allows the composer to create a series of musical states, each with a different probability of leading to the next. The *Reflector* (as seen in Figure 11.12) allows the composer to

Advanced Algorithmic Music Systems

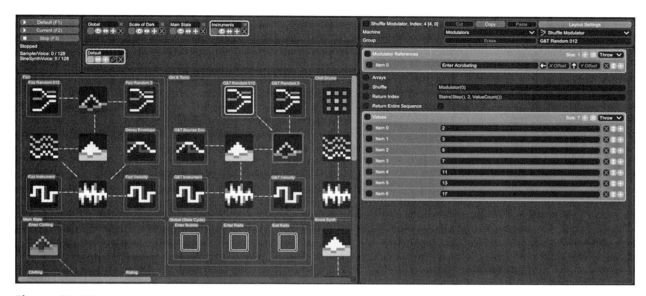

Figure 11.11.
Shuffle Modulator module in *Ondskan*.

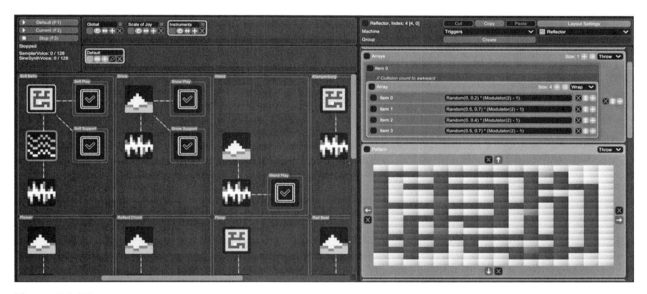

Figure 11.12.
Reflector in *Ondskan*.

create a visual maze that the system will attempt to navigate when triggered by an in-game action. Each time the cursor bumps into a wall, this affects the music.

Most importantly to note, by designing his own custom system, Niklas Nygren has created almost unlimited compositional possibilities. The *Ondskan* system can receive game data from anywhere in the game, and this data can be used to instigate any type of musical change the composer wants by running it through custom-designed manipulation software.

Conclusion

Ondskan is a robust example of the musical possibilities offered by a custom algorithmic solution to music implementation. Using such a system, a composer can easily apply the many compositional techniques we have discussed in this chapter and those preceding it to create incredibly unique and deeply reactive scores.

While some of the examples we have discussed up until this point in the book might seem overwhelming in their own ways, it is important to let these serve as inspiration for working with a team to find a creative way of designing a music system. A system like *Ondskan*, for instance, is impressive in its scope, but most indie games will not require anywhere near this amount of versatility. Rather, determine what type of functions the game you are working on really needs and build a system with a strong foundation, adding more creative tools only if there is time.

Exercises

1. Pick one of your favorite games as a case study. Then, research and analyze its music system. What features does the music design incorporate? How was the music implemented?
2. What changes would you make to the music design to make it more reactive?
3. If the music is procedural, come up with three new ways to affect the music based on game parameters.
4. If the music system is not procedural, create a hypothetical algorithmic music design for the game. How would you implement the music?

Notes

1. *Just Cause 4*, game disc, developed by Avalanche Studios (Tokyo: Square Enix Holdings, 2018).
2. *Uurnog Uurnlimited*, download, developed by Nifflas Games (Stockholm: Raw Fury AB, 2017).
3. Elizabeth Medina-Grey, "Modular Structure and Function in Early 21st-Century Video Game Music" (PhD. diss., Yale University, 2014), 235.
4. In jazz harmony, extensions are degrees in the chord above the octave; in other words, the 9th, 11th, and 13th, all of which can be sharped or flatted to provide more harmonic options and colors.
5. The speed at which the harmonic progression moves; in other words, how many beats each chord lasts.
6. The "swing" feel is rooted in eighth note triplets as opposed to straight eighth notes. When one turns up the swing, the eighth notes are quantized on the rhythmic timeline to the triplet instead of the eighth note.
7. Niklas Nygren (a.k.a. "Nifflas" of Nifflas Games), interview with author, December 3, 2017.
8. For more information about "expressions," refer to Chapter 14.
9. Nygren, interview, 2017.
10. We will learn more about for loops in Chapter 14 about programming. Essentially, they allow the programmer to make the same action be performed repeatedly until a specific point or value is reached.
11. Ibid.
12. Ibid.
13. A mathematical system that randomly or discontinuously moves from one state to another, depending on probabilistic outcomes of the present events. Think of it as climbing a probability tree: the probability of which branch you climb next depends on the branch you are currently on; this is a fun way of creating partial-randomization for navigating musical probabilities.

12

Music in Virtual Reality

VR Technology: A Diverse Playing Field

Virtual reality (VR) is beginning to find traction in the consumer industry, making it an increasingly important platform for game development. The VR gaming market is growing at an incredible pace; various companies, both small and large, have entered this arena, selling a variety of devices aimed at different groups of consumers. As of this writing, some notable examples of VR companies include Oculus (owned by Facebook), Sony, Microsoft, Samsung, HTC, and Google. For example, since releasing the renowned PC-powered Oculus Rift headset in 2016, Oculus has released the Oculus Go, a stand-alone headset aimed at TV-viewing consumers; the Oculus Rift S, an even more advanced, PC-powered iteration of the Oculus Rift; and the Oculus Quest, a stand-alone VR gaming system targeted toward gamers who prefer a portable system. Technology is constantly advancing, and it is likely that by the time you are reading this, many other developers will have entered the market with impressive VR hardware.

What is most important to understand is that VR gaming hardware comes in many different shapes and forms, and because of this, the gaming experience can differ greatly depending on which system the game is made for. With this in mind, composers must approach composing for each technology with an open mind, adjusting their scores to both the technological and experiential requirements of each system. In their 2018 AES talk, "The Role of Audio and Multimodal Integration for New Realities," VR audio specialists Robert Rice and Kedar Shashidhar point out that "these new realities demand an understanding of the connection between audio design and technologies that vary in immersiveness and deliver to variable end-user environments."[1] Here they make a crucial point: audio creators must design with the understanding that a single VR game may be experienced in various ways because of the notable differences in the features of delivery platforms.

Indeed, VR devices vary dramatically in price and performance; likewise, they vary dramatically in the experience they provide players. In contrast to the expensive, PC-powered Oculus Rift, the Google Cardboard VR provides a low-cost option, in 2020 selling for just $15. Constructed of no more than a few pieces of cardboard and some inexpensive lenses, Google Cardboard VR uses a mobile phone as its processor. Similarly, Nintendo (under their Nintendo Labo brand) has created a variety of cheap devices designed to hold the Nintendo Switch. Low-cost VR options like these come with a variety of functional limitations compared to higher-end devices.

As we begin our discussion of VR, it is important to understand that VR is only one facet of *extended reality* (XR), which is a blanket term used to encompass VR, *augmented reality* (AR), and *mixed reality* (MR). AR and MR both serve to bring elements of the digital world into the real world, augmenting it with virtual objects with which the player can interact,[2] while VR transports the player into new realities altogether. XR technologies are changing the way that humans interact with media. While AR and MR have both proven to be useful game delivery methods, VR's complete immersiveness makes it a unique platform for creating new game worlds—and beyond. This new level of immersion also makes VR an important avenue to explore when it comes to composing music.

As we dive deeper into how VR works, it will become quickly apparent that the highly immersive nature of VR experiences provides composers with new challenges. VR fully engulfs players in a new world, occupying their vision, hearing, and—in many cases—even their sense of touch. The player may use a plethora of different controller devices, each one providing a different type of player input for the game. New input devices are also continually being created, making VR experiences increasingly realistic and immersive. The diversity of VR hardware makes designing games and audio for VR quite challenging; however, that same diversity also opens up a world of opportunity when it comes to getting creative with music design.

How VR Works

VR experiences are presented in a variety of ways, but the most common delivery method is through the use of a *head-mounted display* (HMD), also sometimes referred to as a *VR headset* (Figure 12.1). Using an HMD, the player looks into a display, which usually consists of two screens, one for each eye, with angles slightly offset from each other

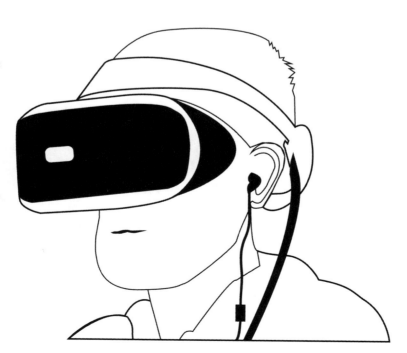

Figure 12.1.
A typical HMR.
Image by Kierstyn Zaykoski and Sean McLeod of New York Institute of Dance & Education, with permission for use in this text.

to account for how each eye sees differently, using auto-focus lenses to adjust constantly to our eye movements in order to provide a realistic binocular perception of depth.

Degrees of Freedom

Many HMDs track only head rotation. These provide *three degrees of freedom* (3DoF). When an HMD is 3DoF, it can track only the player's head orientation: the player can nod his head up and down; rotate his head left-to-right; or tilt his head forward and backward (Figure 12.2). However, the viewing experience is fixed: the player's virtual position in 3D space is not being tracked. In other words, he cannot move forward or backward, getting closer to or further from virtual objects in order to see them from a new perspective. 3DoF devices therefore need only to use rotational tracking since the only player movement being detected is stemming from rotation of the head and/or torso.

Some HMDs use the combination of rotational and positional tracking systems to monitor the player's head and body movements together, and react accordingly. This allows the player to not only look around in 3D space just as she would in real life, but also to actually move around the virtual location, thereby changing her in-game perspective. These HMDs offer *six degrees of freedom* (6DoF). In addition to tracking rotation, 6DoF HMDs use a positional tracking system, meaning players can move in space forward and backward, up and down, and left and right, completely changing the perspective on how people, places, and objects are viewed. Most of the more powerful gaming headsets provide 6DoF, allowing players to explore the game world without being "stuck" in a fixed position; this creates an experience much like that of humans perceiving and interacting with the real world (Figure 12.3).

Processors: PC versus Mobile versus Stand-Alone

VR devices have a wide range of processing power capabilities. Naturally, this has a big impact on what can be accomplished from the standpoint of development. Just as with any other hardware, the more powerful a system is, the fewer limitations creators will encounter in development for that system. In most cases, more power means more potential for a game to support both excellent visual graphics and a robust audio system. PCs are among the most powerful devices currently available for running video games; thus, PC-powered VR devices offer the most freedom for creating graphic-intensive games and VR experiences.

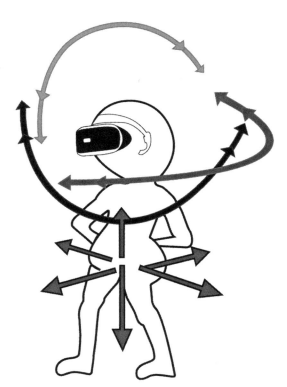

Figure 12.2.
3DoF visualized.
DoF images by Kierstyn Zaykoski and Sean McLeod of New York Institute of Dance & Education, with permission for use in this text.

Advanced Reactive Music Concepts

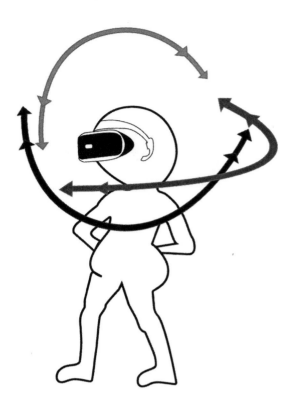

Figure 12.3.
6DoF visualized.

Many HMDs use mobile phones as their processing units. The player initiates the game, then places the mobile phone within the HMD; the player then uses the phone's screen to transmit the game's visuals and the HMD's lenses to create depth perception. Currently, mobile HMDs are most commonly 3DoF, offering a limited experience compared to that afforded by the PC-based HMDs. Nevertheless, some companies now offer positional tracking "add-ons" for phones.

Stand-alone HMDs are those which include an internal processor, meaning they can be played wirelessly without the use of either a mobile phone or a PC. These are fairly appealing because they give the gamer full mobility. While powerful laptops are another option for taking VR systems on the go, many avid users will be hooking their HMDs into stationary desktop computers, which offer the most power for the least cost. Stand-alone HMDs offer an affordable way to play VR away from home, and they also have the added perk of being free of cabling. Naturally, these stand-alone devices cannot compare in terms of power and graphics to *tethered* devices—those which require a connection to an outside processor—but they offer an excellent playing experience nonetheless, with the advantage of cableless movement.

Positional Tracking Systems

Each positional tracking system is also unique. Some are as simple as a single camera watching the player's movements, while others use numerous tracking devices placed throughout the real-world play area. Like the displays themselves, these systems offer a wide range of potential maneuverability, from "very little" to "quite a bit." The HTC Vive, for example, uses Valve's *Lighthouse* positional tracking system, which allows the player to be fully mobile throughout a room, limited only by the cables connecting the HMD to the processor. The HTC Vive Pro ships with the *Lighthouse 2.0* system, an even more advanced iteration of the system, allowing players to move around a play area of up to 10 square meters (about 108 square feet). As of writing this, the Vive Pro costs approximately $735 (679 euros), a price point far beyond the $15 Google Cardboard, which does not, however, offer positional tracking as a feature.

Player Input: Controller Types

Player input in VR is continually advancing as new technology adds more ways for games to receive sensory input. New motion sensor technologies pick up increasingly detailed movements, allowing the player to "touch" and "feel" objects in the virtual world. Currently, the most common devices for input are handheld controllers paired with an HMD. Typically, the player has one in each hand, providing full-arm mobility. Controller input for HMDs will continue to become more and more "realistic" as new technologies are developed. VR chairs and omnidirectional treadmills will also make the physical experience of being in VR incredibly realistic since the player will be able to kinesthetically experience the game. Needless to say, these advancements are not only impressive, but also have broad implications for the way composers will use player input to affect the music and soundtrack playback. While we have discussed many detailed ways to allow player input to interact with our musical scores, these new input methods are providing composers numerous new input styles with which to experiment.

VR Audio Delivery

While some HMDs have built-in headphones, others use built-in speakers instead (though both types are able to create a 3D audio experience). The Oculus Go uses two *spatial audio*[3] speakers on either side of the headset; however, there is also a headphone jack for the player to connect her own pair of headphones. When creating a music design for a game being delivered to this platform, the composer therefore must take into account not only how the music would sound directly out of the built-in speakers, but also that many players might use their own headphones instead. Also, when others are in the room when the player is using the headset, the player might also choose to remove her headphones to hear the game sound through a public speaker. In short, just like the spread of VR devices themselves, the way audio is delivered in VR can become fairly eclectic as well. Because of this large amount of variety, there is no one "right way" of creating music for a VR experience. However, a composer does need to be especially aware of what platforms they are creating for in order to shape their music design most appropriately.

Getting Technical

Because of the diverse array of potential VR devices, every VR game calls for a completely different approach to the audio design. The composer must account for the variety of factors noted in the preceding, all of which may differ significantly from device to device. In many cases, however, a composer may work on a VR game meant for release on multiple platforms. This means that when it comes to mixing the audio, a "one size fits all" approach will often be the best route. The audio must function across different devices with different processors, and it will need to sound good whether coming from any style of speakers or headphones.

With this in mind, it is worth noting that mixing for headphones is somewhat different than for speakers, for a number of reasons. When listening to audio on speakers, one hears not only the direct sound, but also a combination of the sound reflecting and reverberating off the space and surfaces in the environment. Because of these reflections, even if a sound is hard-panned to one speaker, both ears will still hear that sound. Because of the added reflections and room reverberation, the amount of detectable reverb is less

Advanced Reactive Music Concepts

and therefore it sounds subtler when listening outside of headphones. Stereo spatialization is different in headphones as well; each "can" passes its ear a nearly isolated signal, and (depending on the headphones) may also provide complete isolation from the room's natural sonic environment. Panning a sound to one side only in headphones means that only that single ear hears that particular sound.

Additionally, it is important to understand how 3D audio differs from 2D audio. In many cases, speaker setups provide a simple stereo mix. In others, they provide some sort of surround format. In either case, the audio is delivered in a 2D stereo field, meaning the listener can hear *width* and *depth*, but one key element is missing: *height*. When mixing for 3D audio, different sonic elements can also be placed in the *vertical* plane, up and down or top and bottom becoming another continuum, just like left and right in a stereo mix.

Binaural Audio

One way that audio is delivered in VR experiences is called *binaural sound processing*. The concept behind binaural audio is that it essentially mimics the function of the human ears, providing players with audio that spatially sounds similar to sounds in real life. *Head-related transfer functions* (HRTFs) are the algorithms that calculate the representation of how audio is picked up by human ears depending on variables like head shape and ear-to-ear distance. Of course, each human head is different, so the most accurate representation of such a signal would be uniquely personal to each person's physiology; however, a personal HRTF is often an impractical solution since it takes extra time and information to calculate.

As of this writing, there are already numerous plugins that can be integrated directly into DAWs for binaural mixing. One good example is the built-in *Logic Pro X* Binaural Panner (Figure 12.4). The center of the "head" is the center of the diagram in the top section of the plugin. The small green dot is the "puck," which shows the location of a sound in relation to a head. The size of the "space" around the head can be manipulated with the "Size" fader in the bottom left such that moving the puck has a greater effect on perceived distance from the sound source the farther the fader is moved to the right. Playing a track and moving the puck around the sound field will cause the audio the puck represents to move around the head, and, so long as the listener is wearing headphones, the listener will perceive it as such. The "Mode" in the center section of the plugin, set to "Spherical," shows a sphere that visually represents the *vertical* location of the sound.

If you plan to use a binaural plugin in your DAW, then you are most likely going to need to pre-render any binaural panning into your music as is. If you wish to dynamically change binaural panning during gameplay, you will instead need to use a binaural plugin which can be used by the game engine. A variety of such plugins exists, and many of them can be used within game engines like *Unity* and *Unreal*, or inside audio middlewares like *Wwise* and *FMOD*. You should conduct your own research regarding which one will be best for your game, depending on its needs, including which engine you will be using and whether you will be using a middleware solution.

One important distinction to make is this: binaural *sound processing* is an excellent solution for how the audio is delivered to the player's ears in VR; however, binaural

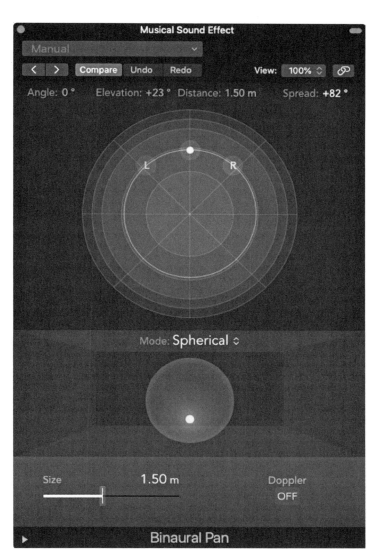

Figure 12.4.
Binaural Panner in *Logic Pro X*.

recording is extremely limiting and is *not* a good recording method for audio intended for VR. Binaural recording involves capturing an audio signal from the perspective of a model human head. It accounts for this single, static representation of how the audio around that specific head moved during that recording. For VR, though, the audio needs to be able to be spatialized such that when players move *their* heads, the audio changes accordingly. Thus, the signal must be localized in 3D space and must be dynamically re-active to the player's position. Binaural recording makes this difficult. A far better solution exists: *ambisonics*.

Ambisonics

An ambisonic microphone, or ambisonic microphone array, picks up sound in a 360-degree sphere—this includes the vertical plane. One of the biggest advantages ambisonics offers is that once sound has been recorded, it can be processed relatively easily and used for many different types of delivery systems, including surround-sound speaker systems or headphones, making it a very versatile recording option.

Ambisonics are broken up into *orders*, which represent how many audio channels they have. The number of channels in a given order is $(n + 1)^2$, where n is the number of the order. Thus, *first-order ambisonics* uses four channels (called "W," "X," "Y," and "Z"). These four channels are able to record the entire 360-degree spectrum of sound using a combination of omni- and bidirectional pickup patterns.[4] First-order ambisonics already provide an excellent recording solution; however, there are also *second-order ambisonics* (using 9 channels), *third-order ambisonics* (using 16 channels), and so on, all the way up to the current highest order, *sixth-order ambisonics* (using 49 channels). Anything beyond first-order is called *higher-order ambisonics*. Higher-order ambisonics are on their way to becoming more frequently used in the industry; however, like most larger, more complex files, they are more costly both in terms of recording equipment and load on the system; thus, in many cases, first-order ambisonics must suffice.

Recording ambisonics, however, may not always be practical, nor is it even necessarily the best solution for creating 3D audio. Thankfully, ambisonic mixing also allows the conversion of traditionally recorded audio into an ambisonic mix. This is done using *ambisonic encoders*, software solutions that allow a stereo mix to be converted into an ambisonic mix. Therefore, audio does not necessarily need to have been originally recorded ambisonically; rather, it is possible to create an ambisonic mix from non-ambisonic material for a VR game if necessary. However, this means that when mixing music, which is generally not recorded ambisonically, into 3D, it must be encoded to properly function in the VR world—especially if it is going to be diegetic.

Designing Music for Virtual Reality

Hardware Questions

With the popularization of VR technologies, an interesting development problem has arisen: the more immersed players are able to become, the more easily an inconsistency of experience can break that immersion. When designing for VR, game developers therefore need to pay extra attention to detail, because players will be doing so, even without necessarily realizing it.

As we saw in our Chapter 5 discussion about codes, the human mind is constantly observing the world around it, allowing successful navigation by making predictions and extrapolations based on past experiences. The mind comes to expect certain patterns based on its knowledge of how the world works. In fictional worlds, new rules can be established, and through repetition, the mind will come to recognize and allow these rules as a new norm. From the psychophysiological standpoint, a chief responsibility of game creators is to ensure that the game world is coherent. This is done by providing the player with the physical, visual, and auditory feedback that the player expects. Therefore, even the smallest inconsistency in the expected physical nature of objects and how they interact with each other can break immersion, reminding the player that he is not actually in this virtual world, but that there is an HMD strapped to his head.

As we have seen throughout this book, audio plays a crucial role in the process of fostering immersion. Just as the player's mind calculates how objects should physically interact, it also has expectations regarding sound. Thus, the soundscapes the audio team designs must be coherent with the visuals of the game. The audio must act and react according to the established rules. Poor physics, bad audio design, and shoddy implementation can all easily break player immersion in VR.

While it would be nice to be able to provide a "one-size-fits-all" statement for how music fits into VR games, this would be misleading and, quite simply, impossible; in fact, one unique but fun challenge of working on VR games is that they, arguably even more so than flat (2D) games, require a unique and subtle approach in designing each soundscape.

When designing the music for a VR game, there are a few important elements the composer should consider. Much like designing for any flat game, the audio design process should begin by ascertaining the hardware to design for. *How powerful is the device? How much resources will be allotted to the music and sound? How is audio delivered to the player's ears? Does the player hear sound through speakers, or headphones? How likely is the player to use his or her own headphones or the built-in HMD audio system?* Furthermore, *will the game be delivered on one specific VR device, or will it be deployed for multiple devices, each with different characteristics?*

Design Questions

After determining the technological possibilities and limitations in a given VR project, the composer can then move on to the conceptual design process. Coming from more traditional areas, VR represents an entirely new world of possibilities when it comes to game design. The heightened level of immersion will offer a challenge to composers in many cases, meaning that they must be extra careful regarding how their music occupies the sonic space. No two games are the same, and despite the uniqueness of VR as a deployment platform, one must always think carefully about how each music design is structured.

VR does present certain new challenges. One such challenge is how sounds are spatialized; in VR, the player will often be hyper-aware of "where" sounds are coming from. Sometimes, this means that a 2D, nondiegetic music score could actually detract from the immersiveness of the virtual experience—in opposition to, say, what the audience experiences in a movie theater, sitting in a fixed seat. The surround-sound mix is tailored to that "sweet spot." The composer designs the audio experience knowing almost exactly how it will sound when music, dialogue, and sound effects are moving around the theater. Nondiegetic music can be mixed knowing that the audience experience will be essentially the same each time the film is viewed.

Of course, the variable nature of head placement at any given moment in VR does mean that the players will experience the sound quite differently than they would for flat games. Predicting exactly whether or not the nondiegetic music will clash with the other sounds in the scene is exponentially more difficult. For example, amidst a loud action sequence already busy in terms of how much of the audio spectrum is occupied, a player might then change the positioning of her head during this gameplay, meaning—as in real life—she would hear certain sounds much more clearly than she would hear others. In other words, the sounds in the stereo field will "move" as the head is moved.

Advanced Reactive Music Concepts

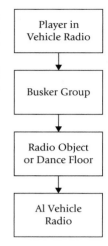

Figure 12.5.
Example of audio prioritization diagram in *Just Cause 4*. Line drawing by Ronny Mraz and Zach Abramson and provided with permission to use in this text.

Audio prioritization is one common way of keeping the soundscape clean and unimpeded by conflicting audio sources—in fact, also employed regularly by flat games. For instance, in *Just Cause 4* (2018),[5] the developers used an audio prioritization system to ensure that the diegetic music did not become overly intrusive. With this system, for example (see Figure 12.5), if the player were in a vehicle listening to the radio, the music from the radio would take top sonic priority. If the player were to exit the vehicle and approach a group of buskers on the street, the sound of the buskers would take top priority, even if the radio in the car were still on. If the player then were to walk up to a group of people listening to a different radio, the sound of that radio would take over the soundscape. This audio prioritization system ensures that whichever sound-creating object is of the most relevance to the player's attentions will be given precedence in the soundscape. ▶(12.1)

As in flat games, audio prioritization is also important for VR. In his 2016 GDC talk, "Composing Music for VR Games: *Adventure Time* Case Study,"[6] composer Erik Desiderio talks about ways in which the music for *Adventure Time: Magic Man's Head Games* (2016)[7] was mixed up or down, depending on what was required in each scene. In general, the music ducks in volume whenever the characters are talking. The audio team also applied a dynamic amplitude system so that the music never interfered with immersion, but supplemented the scenes when possible. As Desiderio explains:

> We would dynamically adjust the volume of the music to achieve different moods. When the character is in the happy part of the forest, the music is louder to create a fun, adventurous feeling. When the player is close to waterfalls and other locations that create sounds, we could change the volume of the music so that the player would hear the waterfall more. This would affect the mood and it was good for the sound effect localization.[8]

Indeed, how the player perceives localization is one of the key differences when working on VR games compared to flat games. Considering the detailed localization achievable using binaural sound techniques, VR music has the potential to be more than simply underscore—even if its aesthetic function *is* as underscore. Without sacrificing the subtlety of a piece of music, it is in fact possible to take nondiegetic music and place it throughout the VR game's soundscape. Done well, musical elements given in-game localizations can give the soundtrack a detailed sense of depth in VR. For instance, rendering a single instrument track and attaching it to a diegetic object is an interesting way of creating inconspicuous, musical interactivity in VR.

When working on *To Be ith Hamlet*—a live, experimental, Shakespeare performance in a virtual space—PHÖZ had the chance to experiment with both musical diegesis and localization. The audience member/player experiences the performances live in real time: actors perform in one location wearing bodysuits with motion sensors that track their movements, and microphones pick up their live vocal performances, and these movements and voices are realized in a VR environment. PHÖZ created a bed of nondiegetic music that remains constant as the player explores the virtual scene; however, it became apparent that a musical opportunity was being missed. It was determined that it would be best to encourage players to interact with the actors' avatars, so a different method was tried: writing each character theme in a specific instrument, and attaching that audio track to the character avatar's head. This created an interesting result: because one of the characters had an extremely tall avatar, his theme's realization gave the music score verticality. Though amplitude changes due to localization were extremely subtle, players would still sense a character's theme grow in volume as they approached that character. This method provided an amount of musical interactivity to this experience—one which encouraged the player to experiment with his proximity to each character—and made the music score more dynamic and personal.

Regarding musical codes and the psychology of how vertical mixing affects the listener, creating a great deal of height in the music has the effect of portraying size, even grandeur. Simply making music "tall" changes how the player will perceive it. "Tallness" can make the music feel bigger than the player, and thus more imposing than it otherwise would be were it mixed only in the stereo field, or lower in the vertical field. Sometimes, placing certain musical elements high in the mix can also make those elements seem "out of reach" since the ear perceives them as "too high to grasp." Conversely, "shortness" can have the opposite effect, making the mix more psychologically "approachable." However, keep in mind that making a sound come from directly *beneath* the player actually takes a way a sense of control; psychologically, this technique makes it feel as if there is no "ground," or that the player is standing on thin air, or could perhaps step into a hole from where the sound is emanating. In short, considering the verticality of the mix provides a new element to consider when crafting the intended psychological response. Themes representing certain characters or locations may benefit from such realization.

Another example of a game incorporating experimentation with localization and placing objects in the diegetic VR world is *Farpoint* (2017).[9] *Farpoint* is a VR science-fiction first-person space shooter that uses the PlayStation Aim controller, shaped like a gun that is held in two hands, thereby aiding player immersion by placing what feels like an actual weapon in the player's hands. The game strives to create a sense of realism to strengthen the immersion, enveloping the player in its alien landscape. Composer Stephen Cox and his collaborators at Sony spent a great deal of time experimenting with where to place the score in the virtual soundscape. Unsurprisingly, they eventually concluded that it was in fact a superior experience when elements of the score were localized in the game's atmosphere:

> In certain levels, the speakers might be built into the wall. Or, into a plane in the world. So, the music is going to change and phase as well as mix depending on where your head is turning. However, music could also be built into the ambiance to where you turn your head and that music is just going to follow you so that it's static and constant. Later in development, we found that it's cooler when the music is a part of the world and actually when those speakers[,] wherever they might be, are just there.[10]

This process worked well for a game like *Farpoint*. However, emotionally driven VR games can sometimes require an added level of subtlety when it comes to writing and implementing music. In games that offer sparse, spacious experiences, the player's heightened awareness of realism can have the effect of making the player even more aware of the presence of a music score, and thus more sensitive to it, causing it to potentially overwhelm the soundscape. Therefore, one approach that many composers are taking when composing for this style of VR game is to craft their scores even more subtly than they would for flat games in the same basic genre. The less busy the music is, the less likely that it will interfere with immersion by taking up too much of the audio spectrum; the less likely it is to clash with other sounds in the scene as players move around and explore the game world freely, experiencing the 3D audio. VR audio specialist Larry Yucheng Chang believes that "... music in VR should be more subtle and be more intimate for the player [than music for flat games]. This is because in VR, each player experience is personal. With the first person perspective, music in VR is all about players' connections to the virtual world. They are the protagonists of the show and the music is built around them."[11] Indeed, this idea of de-emphasizing music is supported by the fact that in many cases, the human mind actually prioritizes visual stimuli over auditory stimuli. Composer and sound designer Jean-Luc Sinclair discusses the "visual dominance concept": that if sound and visuals conflict, the brain will default to trusting the visuals.[12]

Still, for the right type of VR game, going back to the basic technique of using stereo, nondiegetic musical underscore can provide a wonderful gaming experience in VR. In *Moss* (2018),[13] scored by Jason Graves, the player guides a small mouse named Quill through a puzzle-solving action-adventure narrative. The player views each scene as if she is standing above it; for this reason, localization is not a key issue. Nevertheless, Jason Graves's score takes the player on a magical, orchestral journey using traditional underscoring tropes and codes, helping to inform the narrative and enhance the in-game story.

Realism versus Experience

There is still debate about whether VR developers should strive to make experiences as realistic as possible or as entertaining as possible—the real question being whether realism does, in fact, also lead to a more entertaining experience, or whether realism should be sacrificed in order to amplify entertainment value. To quote Sinclair, "We are not simulators, we are storytellers."[14] This argument did not originate with VR; indeed, flat-screen games also beg the question of whether we should focus on realism or entertainment. As we discussed in Chapter 7 about music versus sound design, realistic games tend to shoot for sonic realism as well. Even in fictional environments, audio teams strive to achieve a result similar to what one would hypothetically experience in that fictional reality.

With regard to music, one could argue that a fully nondiegetic score lends itself better to realism because it serves as no more than underscore, a completely separate element from the reality of the game world itself; however, games like *Farpoint* demonstrate that even in first-person immersive games, the score can be integrated with partially diegetic elements localized throughout the game world. When done cleverly, this produces an added sense of depth and interactivity to the music, only adding to the immersiveness of the experience.

To again quote Robert Rice, "It comes down to the design, the intent, the intention. It varies depending on the technology, content, end user environment, the story or the puzzle, and the game mechanics. Most importantly, it's about the creative vision. If the music is interfering with that, it's a problem that can be solved."[15]

Conclusion

These examples illustrate only the tip of the iceberg when it comes to musical experimentation with VR as a medium. It will be up to you to continue to discover new ways to implement music and musical sound effects for virtual experiences. As we have seen, placing music diegetically is not always the most effective choice; when doing so yourself, you should be careful to listen to how player movement affects the music, noting any potential problem areas with localization and how the diegetic elements fit together with the nondiegetic elements as the player moves around. As with any project, the needs of each specific VR game will dictate the appropriate answer.

VR experiences will also continue to differ depending on which VR headset is being used and what audio monitoring setup the player is using. When you design your VR music, therefore, remember to consider who your target audience is and which technologies they are most likely to be using. Ultimately, VR will often require more subtlety in scoring than previous gaming platforms did, both in the music composition itself and in the way it is implemented.

VR is an altogether new platform for us to experiment with. The rules for how to implement audio are still being written. As composers, we now have the opportunity to create verticality in our scores, placing instruments or layers above or below players instead of just surrounding them. This gives our music an entirely different type of vertical depth. It also has different psychological implications on how the player will perceive our music.

Binaural audio provides a highly realistic hearing experience for VR games, while ambisonics offer us an excellent way of recording and mixing 360-degree audio. It allows us to turn our heads in any direction, and the recorded signal will adjust to these spatial changes, creating a very close representation of how we hear in the real world.

All things considered, VR is still a new paradigm. It is not only we composers who are still experimenting, but game designers and developers as well. As designers continue to find new ways of immersing the player in VR, we will be tasked with doing the same with our VR scores; so, never be afraid to experiment, and you too will help us collectively move the VR platform forward!

Exercises

1. Track down a location where you can experiment with playing VR games. If you or a friend has a VR headset, great!
2. Listen carefully to the VR soundscape as you play the game. Try turning your head in different directions as you play, listening to how the sound reacts. Do you notice any verticality in the mix? How does the reverb sound to you in each space?
3. Note three instances in which you felt a sound could have been improved in quality or implementation. Why was this sound ineffective, and how would you recommend improving it?

4. Note three occasions on which a sound felt extremely "realistic" or, at least, appropriate to the game world. How was it localized? How did it sound as you moved your head around the space?
5. Is there any audio prioritization occurring? Do you notice any audio elements being lowered in volume to make space for others?
6. Is the music diegetic, nondiegetic, or mixed? Did you notice any instances in which the music had diegetic elements placed throughout the soundscape?

Notes

1. Robert Rice and Kedar Shashidhar, "The Role of Audio and Multimodal Integration for New Realities," AES New York 2018 (convention lecture), Jacob K. Javits Convention Center, New York, August 20–22, 2018.
2. To further clarify, VR requires players to wear headsets that completely envelope their fields of vision. Conversely, AR takes digital content and makes it a part of reality; the blockbuster success of *Pokémon Go* (2016) brought AR to the forefront of mobile gaming, inspiring players across the world to interact with reality in an entirely new way (by placing virtual Pokémon "into" the real world). MR lies in-between AR and VR; in MR, elements of a fictional world are placed into the real world, creating a "mixed" environment. In other words, the player's experience is neither completely in the real world nor completely in a fictional one, but rather in a blend of the two.
3. Audio that simulates a real-life hearing experience, adding a vertical dimension to the stereo field.
4. Every microphone has a *pickup pattern*, the shape that defines the directions from which it picks up a given audio signal. To offer a very simplistic distinction, a *cardioid* pickup pattern picks up sound from the front, while a *bidirectional* pickup pattern picks up sound from both the front and back of the microphone.
5. *Just Cause 4*, game disc, developed by Avalanche Studios (Tokyo: Square Enix, 2018).
6. Erik Desiderio, "Composing Music for VR Games: *Adventure Time* Case Study," Game Developers Conference 2016 (convention lecture), Moscone Convention Center, San Francisco, CA, March 16, 2016.
7. *Adventure Time: Magic Man's Head Games*, download, developed by Turbo Button (Atlanta, GA: Cartoon Network, 2016).
8. Desiderio, "Composing Music for VR Games: *Adventure Time* Case Study."
9. *Farpoint*, game disc, developed by Impulse Gear (San Mateo, CA: Sony Interactive Entertainment, 2017).
10. Krista Grace, "Music and VR: Interview with Farpoint PlayStation VR Composer Stephen Co," *Edgy Universe* (blog), May 27, 2017, edgy.app/farpoint-PlayStation-vr-stephen-cox-interview.
11. Larry Yucheng Chang (VR audio specialist), interview with author, April 17, 2019.
12. Jean-Luc Sinclair, "Immersion, Audio, and Virtual Reality," The XR Date: Música (convention lecture), April 4, 2019, Espacio Fundación Telefónica Madrid, Madrid, https://youtu.be/kGcNQh9SgEo.
13. *Moss*, game disc, developed by Polyarc (Seattle, WA: Polyarc, 2018).
14. Sinclair, "Immersion, Audio, and Virtual Reality."
15. Robert Rice (VR audio specialist), interview with author, May 2, 2019.

PART IV
Implementing the Score

13

Game Engines and Implementing Music in *Unity*

Introduction

It should go without saying that the process of creating a video game involves a great deal of programming. One of the most complex parts of the programming process is the creation of the game physics which determine how objects in the game world interact with each other. Coding the physics for even a simple game like *Pong* (1972)[1] would involve complex thinking and problem-solving skills to create from scratch—after all, it must be determined how the ball sprite would behave when striking a paddle sprite, involving a variety of different potential angles of incidence, velocities with which it might be struck, what direction it might move in, and how quickly it will "respond" toward any direction. The boundaries of the game—how the ball will react depending on which "wall" it hits—must also be accounted for. As the complexity of a game increases in terms of how many different types of objects, materials, characters, and so on, may exist within a game world, more calculation needs to be taking place at any given moment.

Creating each and every aspect of a game's physics from scratch every time would be an incredibly difficult and impractical method of game creation—especially given the amount of time, energy, and funding it already requires to create games with premade assets! A game engine streamlines the process of game programming by, among other things, including an array of built-in features that allow programmers to not have to undertake the incredibly time-consuming step of developing physics. By using game engines like *Unity* or *Unreal Engine*, programmers can instead begin by using the objects provided by those engines.

Unity is one of the leading modern game engines and has been particularly well-received within the indie game development community because it is both free to use and able to build for multiple popular platforms. It is all but essential for any game composer to have basic *Unity* skills, not only with a general ability to navigate the software, but also an understanding of the different windows that pertain to music and sound implementation. This chapter therefore surveys many of the different audio features available within *Unity* and how they can be used creatively. This will include a discussion of all of the audio components available for implementing sound in *Unity*, as well as its audio mixing functionalities for crafting a solid mix.

The *Unity* Interface

Understanding *Unity*'s interface and how the program functions allows insight not only into *Unity*'s functionality, but also into how modern game engines work in general. Becoming comfortable using a game engine is much like becoming comfortable using a DAW, in that once proficiency is achieved with one, it is considerably easier to learn how to use another because of their inherent commonalities. For example, while we will not be looking in detail at Epic Games' *Unreal Engine 4,* understanding *Unity* will still assist you in understanding how *Unreal* works should you find yourself working on a project in that engine. Because it is the engine of choice for many indie game developers, however, *Unity* is an excellent place to start to understand the concept of the "game engine." It is laid out in a clear and functional manner and includes a plethora of built-in objects that facilitate game development. As we explore *Unity*, we will simultaneously be developing a general understanding of similar components in other engines. So, let us begin our quick overview of *Unity*'s user interface in preparation for diving into the functionality of the software.

The "Hierarchy" window (left of Figure 13.1) displays a list of all of the objects currently in the "Scene" (right). The latter displays an editable version of the game world—game objects can be selected by clicking directly on the objects in the Scene. In Figure 13.1 there is a *main camera*, a directional light, the ground, the player object, and an array of other objects. In *Unity*, a main camera is the way in which the world is viewed in the game. For example, in a first- person game, the main camera might be attached to the player object so that the camera's view follows that of the player as he moves. In the "roll- a- ball" game shown in Figure 13.1, the main camera is set at a third- person view, following behind the player (in this case, a white ball) at a short distance. In many other games, the camera may remain completely stationary, displaying a specific view of the scene.

Camera placement is a key concept for the game composer to understand. After all, the implications of different camera placement options will affect music and sound design decisions, as the "Audio Listener," an object that is essentially the set of ears through which the player hears sounds, is often attached to the camera.

Though not particularly important from the musical perspective, the "Canvas" object is a built-in *Unity* object used to display informational text during the game. In some

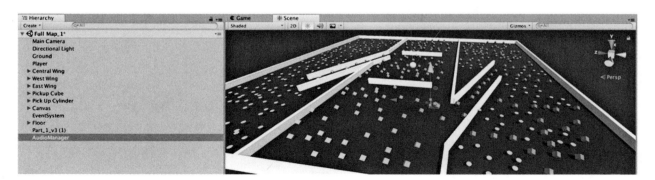

Figure 13.1.
Unity UI. Unity Technologies (2019).

cases, it may be used as a *heads-up display* (HUD) present on the screen at all times; in others, it may be used to create pop-up menus.

Clicking on any object in the Scene causes that object's properties to appear in the "Inspector" window (Figure 13.2), a special window which displays all of the editable parameters of an object. Technically, everything in *Unity* is considered to be an object; in this case, the "Box Collider" and "Mesh Renderer" are two built-in objects that are attached to the main "Pick Up (15)" object.

The "Project" window (Figure 13.3) displays the hierarchical folder structure and files on the hard drive. From here, files and folders from across the computer system can be *dragged and dropped*[2] to their appropriate locations, providing a convenient way to avoid having to leave the *Unity* UI to manage the game's file system.

Another important window is the "Console" window (Figure 13.4). This window provides real time computational feedback as the game is running, allowing the detection of errors and the tools to debug issues that may occur in the game. This window is incredibly useful to all of the game development teams, including the audio team. For example, if a sound is not playing, by running the game, the Console may be used to see if an error occurs when the sound is supposed to play. If there is indeed an error, the Console is likely to be able to pinpoint from which script the error is originating, perhaps even the exact line of code. If no "obvious" error is occurring, the Console can be used instead to debug, stepping through the code line by line to find possible places where a stray piece of code could be causing the unintended behavior and to

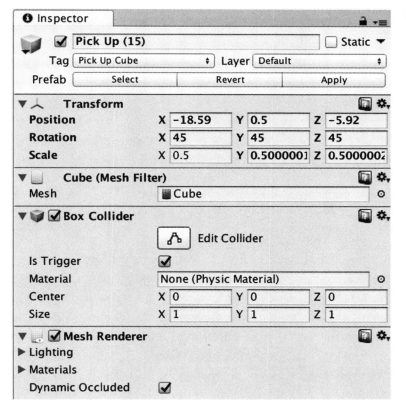

Figure 13.2.
Unity Inspector window.

Implementing the Score

Figure 13.3.
Unity Project window.

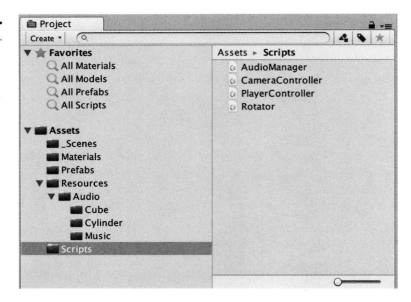

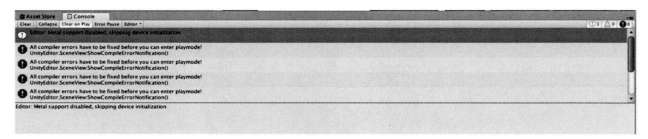

Figure 13.4.
Unity Console window.

print a message illuminating the problem. For example, in a game, a sound should play whenever a ball hits a "pickup cube." Let us use *pseudocode*, a method of writing code in plain English for the purpose of understanding it better. The code reads:

```
If ball collides with pickup cube
    Play collision sound
```

But, here, the sound is not playing. So, to debug, we might change the code to read:

```
If ball collides with pickup cube
    Print "Ball has collided with cube!" to the console
```

If the Console does not print this message, then the problem must be originating in the collision code—in other words, the collision is not taking place when it should. However, if the console *does* print this message, then there may be something wrong with the function that tells the sound when and how to play. There is likely an error somewhere in our coding logic. Either way, the Console's problem-solving and debugging functionality has allowed a look into the inner workings of the code.

Finally, there is the "Audio Mixer" window (Figure 13.5). The "Audio Mixer" is *Unity*'s built-in window for managing audio levels, signal flow, and digital effects. The inclusion of this system in *Unity*'s engine is a huge benefit to the audio community. It

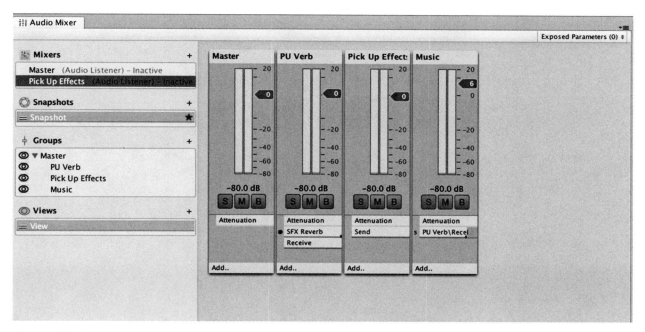

Figure 13.5.
Unity Audio Mixer window.

allows us to create full mixes right within *Unity*, as well as perform other important audio tasks, and even gives us a visual platform with an appearance we are used to from the audio perspective.

Audio Mixer "Snapshots" are essentially like the presets found in audio plugins that allow us to recall various settings: they save the current states of all effects and levels so that they can be returned to easily. This means we can have different mix levels for different scenarios and transition between them as needed. In the case of *Where Shadows Slumber*, we (the PHÖZ team) gave each level its own Snapshot, which would load with the other assets of that level. Using Snapshots allowed us to easily change the audio mix depending on the needs of the environment. For example, World 4, Level 1, begins in the lower part of a city, a sand-covered slum where the destitute inhabitants of the city live. The environment of Level 1 is starkly different from that of Level 4, which takes place in the gardens of the king's palace, with a visual aesthetic reflecting great wealth and extravagance. In the following figures, notice the series of Snapshots on the left, reflecting the different levels of World 4: Figure 13.6 shows the mixer settings for Level 1, while Figure 13.7 reflects those for Level 4. In Level 1, to reflect the sounds of the slums, PHÖZ chose to include bulls and insects as part of the atmosphere. However, at the palace those sounds are turned off. While it is not shown here, Level 4's snapshot also brings down the volume of the wind to "zero" and turns up the volume of a fountain of water to be audible.

Basic Implementation

In order to properly implement audio in *Unity*, it is often necessary to use code. Code is written in a *script*, allowing access to audio objects within the code and controlling how they interact with different objects within the game, much like in our earlier collision

Implementing the Score

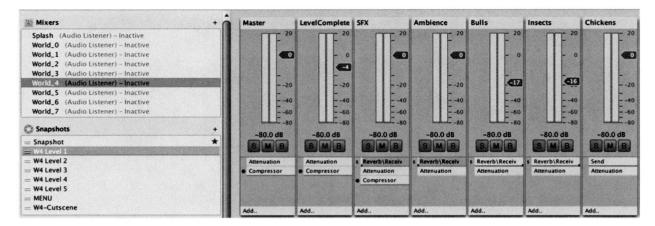

Figure 13.6.
Unity Audio Mixer window for World 4, Level 1, of *Where Shadows Slumber*.

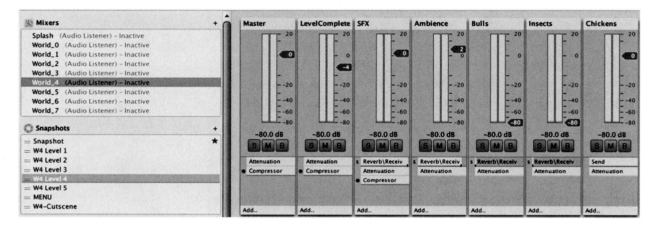

Figure 13.7.
Unity Audio Mixer window for World 4, Level 4, of *Where Shadows Slumber*.

example. As here we are more concerned about the functionality of the code than its literal syntax, we will use pseudocode throughout the following examples to better visualize how *Unity* functions and communicates with scripts.

Thankfully, *Unity* provides a number of useful tools that allow the composer to create basic, well-functioning soundscapes without having to do a great deal of programming. Nevertheless, understanding how audio objects will communicate with the game's programming is an essential skill of any competent music designer.

The Connection between Objects and Code

While the development environment within *Unity* provides us with many wonderful tools, these tools come with certain limitations. Everything in *Unity* is considered to be an object, and thankfully, most objects can be referenced in code. This gives us the power to write our own sets of instructions for how to use objects in our game projects. While we will explore coding in more detail in Chapter 14, let's continue by learning some basic terms used in programming and how they connect to the *Unity* interface.

A *variable* is a small piece of memory in your computer allocated to store certain types of data, which are commonly *integers* (whole numbers), *floats* (decimal numbers), and *strings* (text or sentences), among others. When you *declare* a variable, you are creating an empty placeholder for data or an object of a specific type. If you declare it as a *public* object, this means that you have set its *scope* so that it can be accessed outside of that specific script; whereas if you were to set it to *private*, it would be accessible only within the given scope.

Declaring a variable publicly has an interesting effect when working in *Unity*: it *exposes* the variable in the engine itself, meaning it becomes actually visible in the inspector and you can now access it within the software. Being able to see it in *Unity* allows you to use drag and drop to *associate* a file of some sort with that variable. If, for example, you declare a variable "AudioClip x," you have created a box, or piece of memory, to hold an audio clip (literally, an audio file). Until you have associated an audio clip with that box, it is simply an empty audio clip holder.

Let us see a pseudocode example in which we will declare the existence of a container for a publicly accessible audio clip, then associate an actual audio file with that clip.

```
Script: AudioManager
  A public AudioClip object named "Music" exists;
  Wow, that was easy!
```

Unity, having compiled the code upon saving, now acknowledges that a piece of memory for holding an `AudioClip` exists; this becomes visible in the Inspector (Figure 13.8). However, there is still no file associated with this audio clip; only the container (the reserved memory) exists. Until the container is told which audio file goes inside, it will be empty. Thus, if at this point we were to tell *Unity* to play "Music," we would not hear anything, because *Unity* would find nothing in the audio clip container—there is no file there to play.

To hear a sound, we must place an `AudioClip` in the container. While it is possible to do this with code, it may be easier to instead go to the Project Window and simply drag and drop the `AudioClip` into its proper slot in the Inspector (Figure 13.9). Now, this script can be referenced in association with the proper `AudioClip`.

Now that the audio container has an audio file inside, *Unity* knows exactly which clip to play. However, if we were to click "play" at this point, we would hear nothing! Why? Because the audio container has no speakers—no way for the audio to be heard. It must therefore be attached to a set of speakers so that it will be audible.

Audio Sources

An *Audio Source* is an object that can play audio in the game. In essence it is a virtual speaker existing within the game world. Many game engines beyond *Unity* function similarly, so understanding the principle of how an Audio Source is used will be invaluable in integrating audio in games and understanding "where" the sound and music "come from."

Implementing the Score

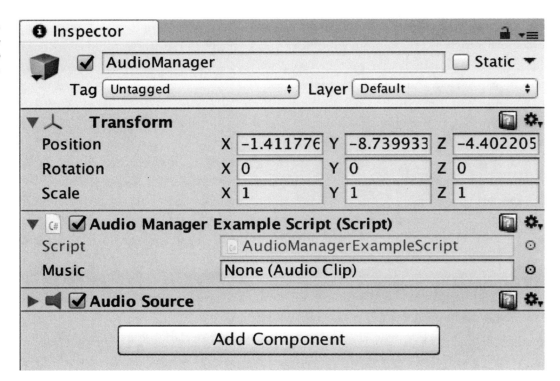

Figure 13.8.
Unity Inspector window with empty AudioClip container.

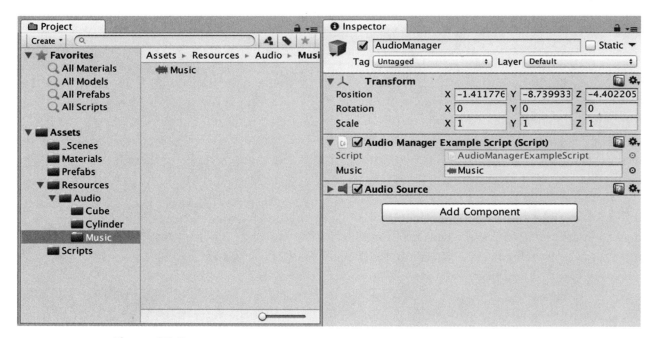

Figure 13.9.
Unity Inspector window with AudioClip.

Similarly to how we told *Unity* that an `AudioClip` object existed, we can do the same for an `AudioSource`:

```
Script: AudioManager
A public Audio Source object named "MusicSource"
exists;
```

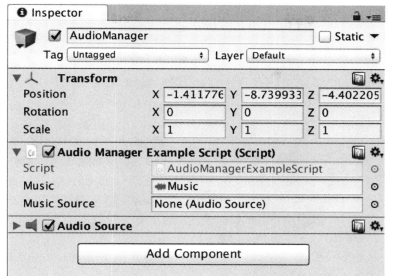

Figure 13.10.
Unity Inspector window with empty AudioSource container.

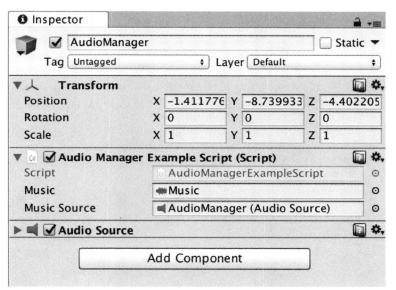

Figure 13.11.
Unity Inspector window with AudioSource.

Just as an empty box appeared for the `AudioClip` once it was declared, an empty box appears for the `AudioSource` (Figure 13.10). *Unity* is now aware that a speaker will be used and referred to by the script. However, it still needs to be told *which* speaker to use. In order to create an `AudioSource`, therefore, we would simply click the large "Add Component" button in the Inspector, search for "Audio Source," and follow the on-screen instructions. An `AudioSource` would then appear on the object in the Inspector. In order to properly associate this "speaker-holding" box with an `AudioSource`, drag and drop the desired `AudioSource` into the box, in the same manner as the `AudioClip` had been dragged in earlier (Figure 13.11).

Finally, when we tell our audio clip to play through "Music Source," it will know which clip to play, and which speaker to play it through!

Audio Source Settings

Unity's Audio Sources provide a number of built-in parameters for properly setting up audio within a game. While we cannot cover them all here, this section will survey some of the most important and most commonly used features of an Audio Source (as in Figure 13.12).

- *AudioClip*: This box shows the `AudioClip` currently assigned to play out of the Audio Source. A single set of speakers can play a number of different clips over time. As the game is played, the associated `AudioClip` might change if the Audio Source is tasked with playing different files. However, the Audio Source can only play one `AudioClip` file at a time; in order to layer them, multiple audio sources will be needed.
- *Output*: This setting determines which "group" or audio track the audio is routed through in the mixer. For example, there may well be a specific "group" with its own fader through which all of the music will be routed. By selecting this "group" as the Output of the Audio Source, the volume of the Audio Source can be controlled through the audio mixer, which, just as in a DAW, can also be used to add DSP effects.
- *Play On Awake*: Checking this box will set the Audio Source to trigger as soon as the game starts. In most cases, this setting should be unchecked, allowing the Audio Source to play via script instead. However, if the music ought to begin playing right away, keep Play On Awake set to "On."
- *Loop*: Checking this box will loop the `AudioClip` indefinitely. This box should be checked if, say, the `AudioClip` is looping music or an ambience track.
- *Volume*: In *Unity*, volume is set from 0 to 1, the latter being an `AudioClip`'s full amplitude. Remember, depending on how the Scene is configured, some control over the volume may be possible through the audio mixer as well.
- *Pitch*: This control sits on a sliding scale, with 1 being an `AudioClip`'s original pitch. Sliding the fader to a value below 1 lowers the pitch, while sliding it to a value above 1 raises the pitch.
- *Spatial Blend*: This control will determine the level of diegesis. If Spatial Blend is set to be 100 percent two-dimensional, the Audio Source will play throughout the entire stereo field—in other words, it will not be directly associated with a specific place or source in the game world. Therefore, nothing taking place within the game world will affect that sound's spatialization; a perfect setting for nondiegetic music. An Audio Source purposed to play music is therefore most commonly set to be two-dimensional, since that setting allows the music to function nondiegetically regardless of what is or is not happening in the game. If the Audio Source is set to be 100 percent three-dimensional, on the other hand, it will be completely diegetic; its placement in the game will affect the spatialization of the sound drastically. We might set it in the middle if we want the sound to always be heard in the background, but we also want to give it an element of 3D localization, its volume growing or diminishing slightly depending on the player's distance from it.

- *Reverb Zone Mix*: This is an area within the game world to which a specific reverb is set within *Unity*. This setting allows sound designers to assign certain levels of a particular reverb to physical areas of the game. As soon as the player enters a reverb zone, a specific reverb is applied. The reverb zone mix fader allows us to control to what extent we wish the reverb to affect a specific audio source's output.
- *3D Sound Settings*: These apply to any sound that is not set to be 100 percent two-dimensional on the Spatial Blend control. They determine how the player will interact with a sound's spatialization within the game. The following five bullet points are examples of 3D sound settings:
  o *Doppler Level*: This controls the amount of doppler effect applied to the sound as it moves in relative space.
  o *Spread*: 0 is fully mono, while 1 is fully stereo.
  o *Volume Rolloff*: This determines what style of attenuation is used when moving closer to or further from a sound. In other words, it determines the rate at which the sound grows louder as the player approaches it.

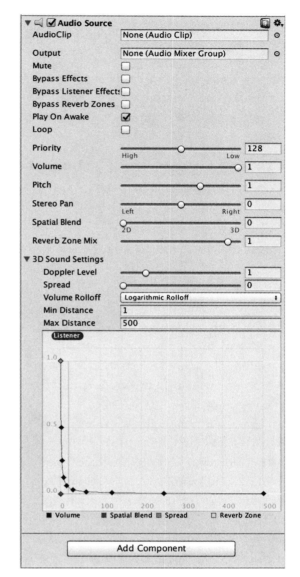

Figure 13.12.
Unity AudioSource parameters window.

  Linear rolloff will give a very steady increase in volume when approaching a sound. Logarithmic is somewhat more exponential and is closer to how the human ear hears naturally. Thus, any Audio Source defaults to a logarithmic rolloff.
  o *Min Distance*: This determines the minimum distance from which a sound can be heard at its full amplitude. Picture each audio source having two invisible bubbles around it, one inside the other. If the minimum distance is five units (as games use "units," rather than physical measurements like "feet" or "meters"), then once the player is within the smaller bubble's five-unit radius, the player will hear the full amplitude of that sound.
  o *Max Distance*: This is the "larger bubble"—the furthest in-game distance from which a sound can be heard. Once the player leaves this larger bubble, she will no longer hear the sound at all. Once she enters the bubble, that sound's amplitude will first initiate, then begin to increase, until the player enters the minimum distance (at which point, of course, she will hear the sound at full amplitude).

As we can see, *Unity* provides composers with a great deal of built-in tools for setting up audio and music within itself. Many of these tools and settings can be utilized cleverly without the use of scripts to create a dynamic audio experience for the player; however, this functionality can be enhanced by using even basic scripts.

In the next section, we will explore some creative uses of *Unity*'s audio engine that allow the creation of interesting, musical soundscapes.

Creative Uses of *Unity*'s Audio Functionality: *Where Shadows Slumber*

Designing an Audio Randomization Object

When designing the soundscape for *Where Shadows Slumber* (2018),[3] we (the PHÖZ team) wanted the environment to feel "alive." The many different creatures in a real environment sing, coo, and groan as they are compelled to; they traverse the landscape, causing branches to crunch under their feet; they growl if they are angry—all in all, producing a natural sonic randomness. In order to emulate a real environment, we needed to be able to randomize a series of elements to create a cohesive sonic atmosphere that assisted in worldbuilding: pitches; amplitudes; placements within the stereo field; and even the time intervals at which these sounds would play. All of these needed to be randomized—but also controlled.

After much thought, we realized that this would be a case in which *Unity*'s built-in features alone could not offer us the functionality we required, but by using a combination of scripts and the interface provided, we would have the freedom to create our own randomization system. We would model it in such a way that its functionality could be used to mimic a real atmosphere. Our final result was a game object we called the "FXMachine." This single object proved to be incredibly useful throughout the game, and is a good example of a successfully designed reusable game object (an object that is useful for a variety of situations or levels and can be used multiple times). In fact, each World in the game has its own version of the FXMachine. Using only a few lines of code, we were able to expose (make variables public and thus visible in the inspector) a variety of variables so that we could easily experiment with them, changing the degrees of "randomness" for each sound and its FXMachine settings.

In Figure 13.13, we have set up the FXMachine to handle bird sounds for World 0. We had wanted the sound of the birds to have a musical quality, much like the sound of real birds does, but in order for it not to be repetitive, we had to utilize the FXMachine's randomization capabilities. Therefore, rather than creating a single loopable "birds" layer of ambience, we ended up creating a series of different bird sounds bounced out into short audio files. Each one had a different melodic quality, was either synthesized or created with flute and violin effects, and carried a certain tune. All of these short audio files were then placed into an assets folder, where they are then selected randomly by the FXMachine each time a sound plays.

"Min Volume" is the minimum amplitude level at which a sound could play, while the "Max Volume" is the maximum; the "Min Pitch" and "Max Pitch" settings function similarly. We also had control over where in the stereo field each sound could play by

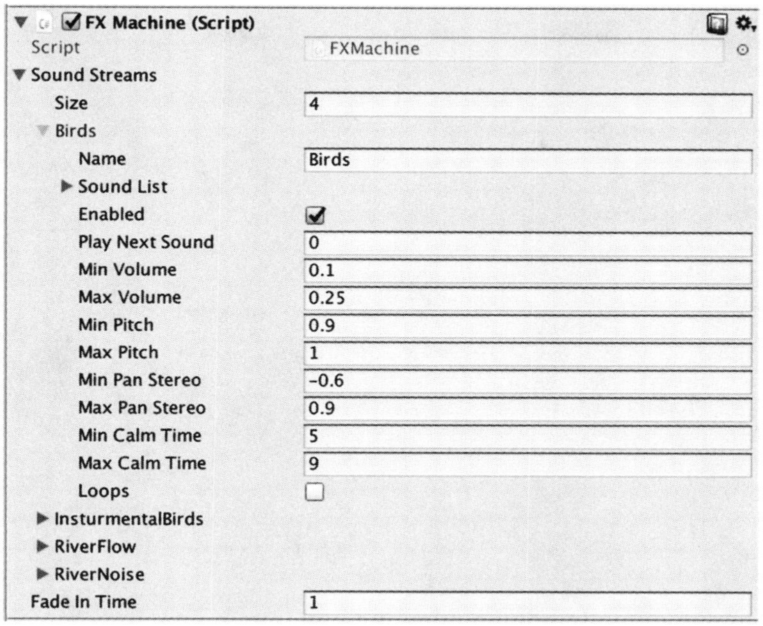

Figure 13.13.
PHÖZ's "FX Machine" parameters window for *Where Shadows Slumber* (2018) in *Unity*.
Where Shadows Slumber, mobile, developed by Game Revenant (New York: Game Revenant, 2018).

randomizing the audio source's stereo pan with "Min Pan Stereo" and "Max Pan Stereo." Finally, the "Min Calm Time" is the smallest interval of time that would pass before a sound could be triggered again, while the "Max Calm Time" was the maximum. By the addition of all of these controls, we were able to experiment with the settings as we played through the game, changing them until the soundscape felt like a living, breathing jungle, achieving a certain amount of realism. It was important to tweak these settings in a "human" way to make sure that none of the randomization or repetition was actually detectable.

It is important to note that if PHÖZ had not been part of the design process from an early stage, we might not have been able to design this FXMachine object. Joining the team when a full year of development remained gave us the freedom to design our audio system in great detail. Without our FXMachine, we would likely have had to bounce out full ambience loops instead, which would have left the game with a far less interesting and dynamic atmosphere.

Designing a Dynamic Music System

The puzzles in *Where Shadows Slumber* tend to be fairly difficult for the players, especially before he gets used to the game mechanic of using a shadow to reveal hidden objects and parts of the level. At times, the player might find himself in a situation where he might be on the right track to solving a puzzle without even realizing it. Having experienced this issue ourselves in other games, we decided to use the music as a sonic tool for player encouragement, signaling to the player when he has discovered the correct course of action. We did this by creating what we called "Puzzle Solve" cues: layers of music that enter on top of the base layer, in sequential order, as each piece of the puzzle is solved.

Since we knew that we would use this system across many of the game's levels, we designed a Music Manager to handle all of the game's music needs. Much like the

Implementing the Score

Figure 13.14.
Music Manager for *Where Shadows Slumber* (2018) in *Unity*.

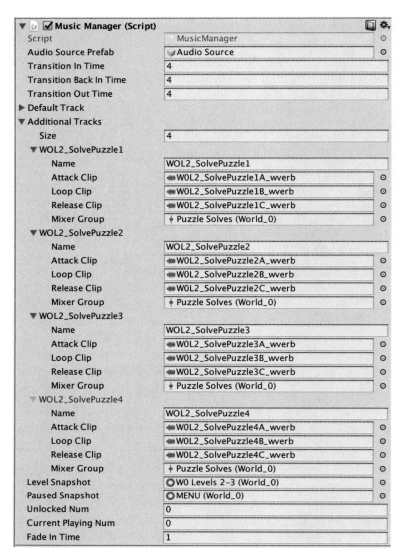

FXMachine was designed to be reused across multiple levels, the Music Manager would also be a reusable object with built-in functions for controlling the music throughout the game. As you can see in Figure 13.14, each level has its own "Default Track." This track fades in as soon as a level is loaded. We also needed a way to introduce our Puzzle Solve layers. We created *triggers* (certain events that take place in the game which, when activated, generate a response) for each specific layer to enter.

For each layer, we created three separate audio files: an attack clip, a loop clip, and a release clip. The attack clip plays as soon as a layer is triggered, then transitions seamlessly into the loop clip. The loop clip continues until the very end of the level, at which point the release clip is triggered as the audio fades out. Using this method allowed us to have a musical "attack" when a particular Puzzle Solve layer is first triggered, giving the player a sense of the layer's entrance. If this attack were heard each time the loop clip repeated, however, the loops would be far too noticeable. Thus, to account for this issue, we created a separate loop clip that loops seamlessly without any noticeable initial attack point.

In *Where Shadows Slumber*, each level contains its own instance of the Music Manager, which handles the transitions to that level's respective Snapshot in order to ensure all of the mix levels are set up correctly. So the Music Manager for each level did not have to be re-created for each level; we created an object with an Audio Source attached to it that could be instantiated whenever that Audio Source was required. This way, we could create and destroy Audio Sources as needed so that there are never too few and never more than the game actually needs at any given time as it is running. In this way, we could ensure that the game would not use unnecessary resources.

The Music Manager also handles the transition to the "Pause Snapshot." We decided that, since this game is for mobile platforms, it would be best to simply have a silent pause menu in case users were to pause the game to avoid producing sound from their phone's speakers. When the menu is opened, the transition to the pause snapshot is triggered and all of the faders go down to zero over the course of a couple of seconds, leaving the game completely soundless (aside from a few UI player feedback sounds).

Though our development and implementation needs surpassed the capabilities of *Unity*'s built-in tools, as you can see, we were still able to handle all of our audio needs for *Where Shadows Slumber* through some simple applied scripting. Thus we were able to design a reactive, and sometimes interactive, musical soundtrack completely in *Unity*, without having to resort to an audio middleware solution. Beyond reduced complexity, using the game engine's built-in audio features also saved the developers money, as middleware solutions tend to have licensing costs for commercial projects.

Conclusion

Unity is the most commonly used game engine among indie developers. Understanding its interface and how to implement audio within it is an invaluable skill for any game composer to have. In this chapter we saw an introduction to how *Unity* works and some creative examples of how to implement audio using its built-in tools and scripting capabilities. Much of this information should prove highly useful not only when venturing further into *Unity*'s functionality, but also when exploring other game engines.

Exercises

1. Write a looping musical track.
2. Download the free version of *Unity 3D* from their website.
3. Open up a new project. Right click in the hierarchy and create a Game Object. Name it `AudioManager`.
4. With your `AudioManager` object highlighted in the hierarchy, go to the Inspector and click "Add Component" to attach an Audio Source to your `AudioManager`.
5. Import your looping musical track from step 1 to a folder in *Unity*'s project window.
6. Assign the file as the clip in your `AudioManager`'s Audio Source.

7. Set the Audio Source to "Play On Awake" and click the "play" button at the top of the window to run the game. You should hear your musical loop playing in the otherwise empty scene.
8. Experiment with the different settings available on your Audio Source to see how they affect the sound of your music.

Notes

1. *Pong*, arcade, developed by Atari (Sunnyvale, CA: Atari, 1972).
2. A method of "grabbing" a file or folder with a pointer, moving it across the screen, and "dropping" it in the proper location.
3. *Where Shadows Slumber*, mobile, developed by Game Revenant (New York: Game Revenant, 2018).

14

Programming Crash Course for Composers

Introduction

Game audio programming is the technique of creating the systems that composers and sound designers use to implement audio into games. This can be as basic as a short script[1] in *Unity* or as complex as creating an entire infrastructure for how sound is managed in a game. The structure and functionality of a game's code often have a large impact on the music design within, so even if you never plan to program by yourself, being able to communicate what the music needs is invaluable during the design process.

Regardless of whether or not you are working for an indie game, AAA, or anything in between, understanding the basics of programming and how it relates to audio is a skill that will set you apart as a composer. This chapter will give you enough knowledge and understanding of audio programming to communicate with your development team and even be able to script some of your own basic systems if needed. We will begin with the core basics of programming, discussing the fundamental building blocks of a programming language, as well as common coding practices. We will then begin diving into the basics of understanding a *Unity* script written in C# ("C Sharp").[2] Finally, we will learn about audio programming, starting with playback of a simple sound, and moving onto building an audio manager that could handle the majority of the audio-related functions of a game. This chapter will leave you with a foundation that you can build upon to keep learning programming beyond the concepts discussed within its pages.

Programming Basics

Programming is how humans give computers instructions.

You might wonder how much math you actually need to understand in order to program. The answer to that question is that programming is based more upon logic than math. While math, of course, is required for the computer to carry out certain calculations, only a basic understanding of mathematical concepts is needed to program at even an intermediate level.

Programmers create useful code by breaking apart a task into its smallest individual steps. Every time a computer carries out an action, it is actually following a set of detailed instructions. For example, when a user clicks the mouse pointer on an application icon and the application opens, somewhere in the computer's operating system code, the programmer gave this instruction: `If application icon is clicked, open program`.[3]

Instructions are made of *statements*.

Statements are sentences that tell the computer what to do. They use words, numbers, and punctuation to express a thought; an individual piece of actionable information. The instruction we used in the preceding is an example of a statement.

In programming, statements can look like

`x = 20`

This is called an *assignment statement* because it assigns the value of `20` to `x`.

An *expression* is like an incomplete sentence that *returns*—produces or expresses—a specific value. For example,

`8 - 5`

is an expression that returns the value of 3. Some other examples of expressions are

`6 + x`

and

`x > y`

Statements often contain expressions. For example,

`if x > y, play exploration music.`

Programming Languages

All programming languages are used for the same basic purpose: to tell the computer what to do. Just as in spoken languages, *syntax* can be thought of as the grammar of a programming language; that is, how the words and punctuation are strung together to form complete sentences (or in this case, statements). Each programming language has its own syntax and is therefore written differently. Each language also has its advantages and disadvantages, and this is why programmers favor specific languages for different types of work.

There are two basic categories of programming languages: *high-level* and *low-level*. Low-level languages communicate directly with the hardware of the machine—in fact, the lowest-level language is called "machine code," which uses hexadecimal values to represent the binary the machine uses. A high-level language, on the other hand, is built upon a low-level set of instructions that carry out a great deal of important actions without the programmer ever having to specifically program them. In other words, the advantage of using high-level languages is that they do a lot of tedious work automatically. For example, take memory allocation: creating a variable[4] called `playerScore` using a high-level language, all the programmer has to do is state

```
create a variable called 'playerScore' of type 'integer'
```

The high-level language automatically allocates the proper amount of memory for holding that integer. If the programmer were using a low-level language, she would have to add at least one extra step to

```
create a variable called 'playerScore'
```

like

```
place 'playerScore' in this specific section of memory
with that amount of bits allocated.
```

Furthermore, not only would each such memory-write need to be explicitly stated, but the programmer would have to understand how to best allocate the available memory as the program became more complex. However, the advantage of a low-level language is that it gives the programmer excellent control over the computer's hardware, allowing the programmer to access certain elements of the hardware directly through the code. Using a low-level language requires a great deal of skill and time because it is more complex to write and manage, and thus it is easier to cause problems with bad or sloppy coding practices. Thus, although high-level languages afford a less-granular control, they can save programmers a great deal of time—not just spent programming, but of studying hardware, too. The language we will be using in this chapter, C#, is high-level programming language often used in games crafted in *Unity*.

Speaking of C#, you might have noticed earlier that when naming our variables, the first word is lowercase (e.g., 'playerScore'). When the first word is lowercase, and all words following are uppercase, this is referred to as camelCase. When all words including the first are uppercase, this is called PascalCase. The convention when coding in C# is to use camelCase for variables and function inputs, and to use PascalCase for functions and classes. We will soon learn more about what functions, inputs, and classes are, but now you understand why words are written a certain way.

Object-Oriented Programming and Class Structure

Modern video games are generally built using *object-oriented programming* (OOP) languages. In *Unity*, for example, anything in the game world is considered to be an *object*. The main advantage of using an OOP language is *class structure*. A *class* is a way to encase several characteristics of an object in one group so that the programmer does not have to build similar items over and over from nothing, which would waste both time and memory resources. For example, a class called Car could be used to create different types of cars in a racing game. Every car could have four variables: acceleration, handling, speed, and powerup. In the class would be functions for each characteristic, telling the game that the acceleration variable controls how fast the car can reach its max speed, the handling variable controls how well the car handles turns, the speed variable will be the car's maximum speed capability, and a powerup will temporarily boost one of the other three characteristics.

C# Essentials

- *Variable*: While we did cover the basics in Chapter 13, let us examine variables one more time. A variable is, quite literally, a section of computer memory that is reserved to hold a specific *value*. That value can be an *integer*, a *float* (a number with a decimal value), a *string* (a word or group of words), a *reference* to a game object, or other various types of data. Now, think of a variable as a "box"; that is, if you were to create a variable called `enemy1`, you would, in effect, be creating an empty box called "enemy1." You could then place a *reference* to the actual file (containing all of the information that tells the game how to create the enemy) into that variable. Thus, whenever you subsequently instruct the game to make an `enemy1`, it will use that reference to find the file on your computer and create one in the game based on the specifications.

  Most variables are *declared* at the top of a given script. A variable is *declared* by giving it a name (e.g., `enemy1`) and a *type* (what kind of object or value the variable will hold). For example:

  ```
  int x;
  ```

  With this simple line of code, the computer will reserve a section of memory where the value `x`, whatever integer that may be at any given time, will always be stored.
  Similarly:

  ```
  int playerScore = 0;
  ```

  The computer will now reserve a section of memory for `playerScore` and set its starting value (`int`) to `0`. However, this value will likely increase as the game goes on; this same section of memory will store the player's score as it increases throughout the game. This means that if the player's score becomes `15`, the `playerScore` variable will now hold the value `15` instead of `0`.

- *Function*: A set of instructions that tells the computer to carry out an action. Functions are often used for instances in which the same action must occur multiple times in a game. For example, a function called `PlaySound()` might be tasked with playing a sound through the audio out.

  After carrying out its task, a function generally *returns* a specific value. For example, a simple function that adds two numbers together will *return* the sum of those two numbers. However, a function that should carry out a task without returning any specific value receives a `void` at the beginning. A basic `void` function therefore looks something like this:

  ```
  void PlaySound(AudioClip clip)
  {
    AudioSource.clip = clip;
    AudioSource.Play();
  };
  ```

  The first line creates a function called `PlaySound()` which receives information about which audio file to play each time it is used; `clip` is an *input parameter*, which is information we provide the function each time we call it. By giving the function inputs, we tell the function to *receive* the information we are providing it *right now* and carry out the instructions using that information. Using inputs, we can determine things like which clips will play and which game object they will play on, depending on what we want in each case.

To use a metaphor: when you build a function, you are creating a CD player that accepts a CD and gives it the instruction to automatically begin playing that CD. What if, however, you wanted that CD to begin playing from track 2? Well, since your CD player accepts an input called "track number," you can tell it to play from track 2 instead. With a function, we can do the same.

The following lines create the function `void PlaySound()`:

```
void PlaySound()
{
  AudioSource.Play();
};
```

We have just declared a set of instructions for the "CD player." However, as there is nothing currently between the curly brackets, the "CD player" will simply play whatever clip is currently set in the audio source. If we want this function to play a specific clip, we need to rewrite it like this:

```
void PlaySound(AudioClip clip)
{
  AudioSource.clip = clip;
  AudioSource.Play();
};
```

Now, our function receives an `AudioClip` which we will simply call `clip`. Then, in the *body* of the function (the section between the curly brackets), it sets the audio source's current clip to that clip and then plays it. Let's say we want our function to play the sound of a laser blast, and we have already declared a `laserBlast` clip with our laser blast sound stored in the variable. We can now call our function and ask it to play the `laserBlast` with one simple line of code:

```
PlaySound(laserBlast);
```

Once a function has been declared, it can then be *called* whenever that specific set of instructions needs to be carried out. To *call* a function is to use that function in a particular context. Functions allow us to forgo having to write out the same set of instructions many times. Instead, we can just write one line: the name of our function. When the function is called (i.e., we write its name), it will carry out the instructions we set when we first declared it.

What if we want a function that will play a sound at a specific pitch and volume? This function would require more input parameters and instructions about what to do with each of those parameters. First, our new function `PlaySound()` will look for `AudioClip 'clip'`, the audio file which it should play. Then, it will accept two integer values, which will be called `pitch` and `volume`. In other words, within the function, we first set the `AudioSource`'s `clip` to a specific audio file; then, `pitch and volume` are set to whatever integer values we enter when the function is called. Finally, the `AudioSource` plays. The function would look like this:

```
void PlaySound(AudioClip clip, float pitch, float volume)
{
  AudioSource.clip = clip;
  AudioSource.pitch = pitch;
  AudioSource.volume = volume;
  AudioSource.play(clip);
};
```

- *If* statement: A simple concept, but one absolutely essential to programming logic. At its most basic level, the *if* statement checks if something is *true* or *false*; then, based on the outcome, performs a function. For example, take the pseudocode

  ```
  if ObjectA and ObjectB have collided, play CollisionSound.
  ```

  If `ObjectA` and `ObjectB` have collided, then the answer is `true` and `CollisionSound` will be played. Otherwise, it is `false`. However, there are many cases that require a more detailed assessment of what has occurred in the game, for example:

  ```
  if ObjectA and ObjectB have collided AND they are both
  green, play sound 'greenSound'
  BUT
  if ObjectA and ObjectB have collided and at least one is
  NOT green, play sound 'errorSound'
  ```

  If statements are often used to create these more complex logical structures by using the term `else if` for other situations in which the primary check is untrue. For example:

  ```
  if ObjectA and ObjectB have collided, play sound
  'collisionSound'
  else if they have not collided, play sound 'gameOverSound'
  ```

  In the following *if* statements example, the *comparison operator* == (double equals sign) is used to compare two values to one another and to check if the condition is true. Following our earlier exploration of *state changes* and horizontal sequencing in Chapter 2, in this example, the *if* statement checks whether `state` is indeed currently set to `exploration`.

  ```
  if (state == "exploration") {
    PlayExplorationMusic();
  } else if (state == "stealth") {
    PlayStealthMusic();
  } else if (state == "combat") {
    PlayCombatMusic();
  };
  ```

  As we can see, the humble *if* statement is one of the most critical programming techniques to understand because it can be used to create complex logic structures and instructions.

- *For* loop: another integral part of programming. The *for* loop is particularly useful for any action that needs to be executed repeatedly. Perhaps the player has reached a "Game Over" in a *Mario* game. As the level reloads so the player can make another attempt, a *for* loop might be used to repopulate the level with coins—reading in pseudocode:

`for every coin object with which the player collided` (and thus destroyed), `instantiate` (create) `a new coin object in its place`.

The *for* loop would run until there were no longer any remaining coins to replace, and then it would stop. Here is an example of a *for* loop in C#:

```
for (int i = 0; i < 3; i++)
{
    Instantiate(coin);
}
```

Behind the first semicolon within the parentheses is `int i = 0` which tells the *for* loop to begin the count at 0. Behind the second semicolon is `i < 3` which tells the loop to run so long as the value of `i` is less than 3. Finally, `i++` tells the loop to increment a value of `1` to `i` each time it runs. This loop would therefore run three times, creating three `coin` objects, at which point it will see that `i` is no longer less than 3 and stop looping.

A Very Basic Game Program

In order to begin our dive into audio programming, we first need a game for the audio to be implemented into. Let us therefore now go over the functionality of a very simple game, which will give us a place in which to hone our audio programming skills.

In *Simple Pickup Game*, we begin with two objects sitting on the ground in front of us (Figure 14.1). The white sphere serves as our "Player" object and has a script, "SimplePlayerController," attached to it.

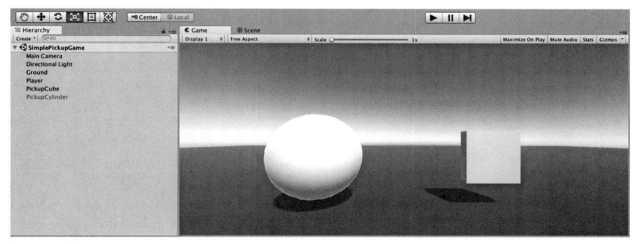

Figure 14.1.
Two objects in *Simple Pickup Game*. Unity Technologies (2019).

Implementing the Score

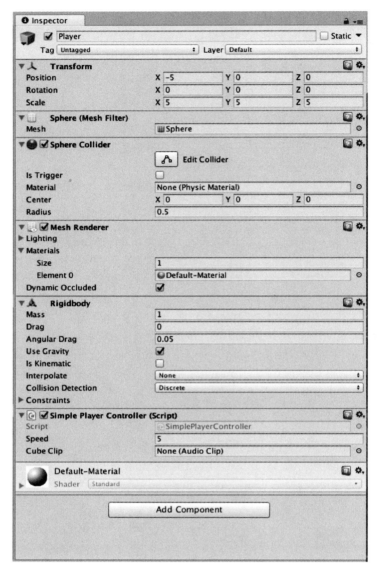

Figure 14.2.
SimplePlayerController with public variables in Simple Pickup Game.

When a variable is *public*, this means it can be accessed outside of its class. In Unity, this also means it becomes visible in the software itself, allowing us to have a visual connection between the code and the GUI (graphical user interface). In Figure 14.2, "SimplePlayerController" has two public variables: `Speed`, which controls the pace at which the sphere will roll when the player moves it, and `CubeClip`, which is a variable to contain the `AudioClip` that will sound when the cube is "picked up" by the player.

A *collider component* (the "Sphere Collider" attached to the `Player` object, seen in the *Unity* Inspector in Figure 14.2) provides an invisible framework for an object's functional shape—it is the "boundary line" of collision, allowing events to be programmed to occur when two objects "touch" each other. In the case of this game, the cube object is currently set to disappear when the player object (the ball) collides with it, giving the impression that the cube has been "picked up." We can then set it so that in addition to the cube disappearing when the objects collide, a sound will play.

```
SimplePlayerController.cs
1   using System.Collections;
2   using System.Collections.Generic;
3   using UnityEngine;
4   using UnityEngine.Audio;
5
6   public class SimplePlayerController : MonoBehaviour {
7
8       public float speed;
9
10      private Rigidbody rb;
11
12      public AudioClip cubeClip;
13
14      void Start () {
15          rb = GetComponent<Rigidbody>();
16      }
17
18      void FixedUpdate ()
19      {
20          float moveHorizontal = Input.GetAxis ("Horizontal");
21          float moveVertical = Input.GetAxis ("Vertical");
22
23          Vector3 movement = new Vector3 (moveHorizontal, 0.0f, moveVertical);
24
25          rb.AddForce(movement * speed);
26      }
27
28      void OnTriggerEnter(Collider other)
29      {
30          other.gameObject.SetActive(false);
31          AudioSource.PlayClipAtPoint(cubeClip, transform.position);
32      }
33  }
```

Figure 14.3.
The code of *Simple Pickup Game* in an IDE. Jon Skinner, *Sublime Text* (3.2.2). Sublime HQ (2019).

Figure 14.3 shows what code might look like in an *IDE* (Integrated Development Environment), a program that makes programming easier by providing a variety of specialized tools and a visually-organized[5] platform in which to code.

Like all game engines, *Unity* has a number of built-in functions, enabling the coding process to move far more quickly than it otherwise would, since developers do not need to create these functions themselves out of nothing. In order to use *Unity*'s built-in functions, *Unity* must know which *libraries* (collections of built-in programming tools) it needs to access in each script. Lines 1–3 appear by default each time a new script is created for *Unity*, as almost every script uses these three libraries. In line 4, `UnityEngine.Audio` tells *Unity* that it will need to access the audio functions *Unity* library in this script.

On line 6 is the declaration of the `SimplePlayerController` class. Everything from line 6 to line 33 will be placed within the two curly brackets { } that define the boundaries of being within this class. In Line 8, a variable is created[6] that will hold a *floating-point number* (a number that uses decimal values) called `speed`. Since `speed` was declared publicly, it will be able to be accessed in the Inspector. In line 10, a `Rigidbody`, which is the component in *Unity* that grants objects the ability to be affected by physics, is declared. As it is named `rb`, it can now be referenced in the script with this name. Finally, on line 12, a public `AudioClip` called `cubeClip` is declared; this will be used in line 31 to play a sound when the player collides with the `PickupCube`.

Unity has a number of predefined functions that are found in many scripts because every *Unity* script has a particular order of events which it follows. These predefined functions are called *Execution Order of Events*. These include the `Start()` function, which is run as soon as the game starts, giving developers a place to add code that needs to run at the beginning of each play-through but not at other times; the `Update()` function, which runs every frame, making it ideal for any variable or function that needs to run repeatedly ("update") as the game is running; and `FixedUpdate()`, which runs every single frame just like `Update()` but is built specifically for physics events.

Going back to our *Simple Pickup Game* example, while `rb` the `Rigidbody` has been declared in line 10, it had not been associated with any specific `Rigidbody` in the game, so it is currently unusable. In line 15, therefore, the built-in `GetComponent()` function of *Unity* tells the game that `rb` is equal to the `Rigidbody` currently attached to the same object as the script. In this case, that game object is the player object, the sphere.

In lines 20–21, two `floats` are created, which hold the horizontal and vertical positions of objects in the game. In line 23, a `Vector3` object (a triple set of coordinates that holds an x-axis, y-axis, and z-axis position) is declared. Here, `y` is set to `0.0f` because the ball should not be able to move up and down the y-axis. The ball's current x and z positions are fed to our `Vector3` (called `movement`) by using the `moveHorizontal` and `moveVertical` variables as defined in lines 20–21. In line 25, the `AddForce()` function, which applies a physical force to an object, uses the direction the ball is moving in multiplied by the public speed variable (a number easy to experiment with in the Inspector).

Finally, in line 28, the script is sufficiently ready to receive audio programming! Remember the public `AudioClip` called `cubeClip` from line 12? This variable is currently empty and has no `clip` associated with it. To add a clip, go into *Unity*, click on the player object in the hierarchy (the section which contains a list of all of our game objects), then go to its Inspector, and drag an `AudioClip` from the Project Window into the `AudioClip` slot. In this scenario, we have dragged the "Csharp" `AudioClip` into the "Cube Clip" public variable container. Now that there is a clip associated with the variable, the script can access it (Figure 14.4).

Remember that a Collider is an invisible "outline" surrounding an object. The shape of the Collider is usually the same shape as the object itself. However, having a Collider does not necessarily mean that an object can be "hit"—for this, *Unity* uses *triggers*: If the Collider is not set to "Is Trigger" then the object will "hit" another; if the Collider is set to "Is Trigger," one object will "pass through" the other (unless otherwise specified), but the collision will still be detected by the engine.

Since in this case we do not want the two objects to physically hit, but rather we want one of them to disappear, we set the `PickupCube` to be "Is Trigger." When the two objects touch, a message will be sent to *Unity* that the player has collided with the Cube, which allows the initialization of the `OnTriggerEnter` function. On line 28, it says that whenever the current object (here, the player, to which the script is attached) collides with the trigger of another object, that object will first be set to be "inactive" (line 30), thereby causing it to disappear from existence in the game. Then, in line 31, the built-in audio function `PlayClipAtPoint()` will play a single `AudioClip` at the specified in-game location. This function is called by first accessing `AudioSource`

Programming Crash Course

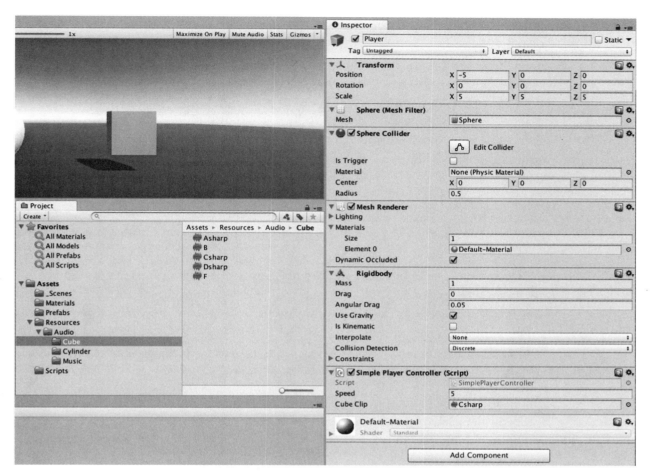

Figure 14.4.
A clip has been associated with the public variable "Cube Clip" in *Simple Pickup Game*.

and telling it to `PlayClipAtPoint`. Then, using the function's input parameters, it is told to use `cubeClip` as the `clip`, and `transform.position` as the `location` (the current location of the object the script is attached to). Now, when the ball collides with the `PickupCube`, the cube will first disappear (per line 32) and then the engine will play `cubeClip` at the position of the player object.

Audio Programming

We have so far created a simple game in which the player controls a rolling sphere. Using the `OnTriggerEnter()` function checks for whenever that sphere touches a `PickupCube`. When this occurs, the `PickupCube` disappears and the `cubeClip` sound is played.

Creating an Audio Manager

So, we have now seen how a very simple game in *Unity* can function. We have integrated a basic audio function by telling the game to play our `cubeClip` at the current

Implementing the Score

position of the player. However, almost any game you might work on will be far more complex than this example and will require an organized system with which to keep track of the audio assets. You might have hundreds (or even thousands) of audio files. You might also have numerous audio functions that you want to be able to reuse, calling them from multiple scripts without having to rewrite them. Because of this, it is an excellent practice to create an audio manager to keep track of both your code and your files as they are loaded into the game and used by the engine. Therefore, let us look at a more complex version of our pickup game example and create an audio manager to manage our audio needs.

In this new version of our pickup game, seen in Figure 14.5, we have two different halves of the level, each containing a different type of pickup object. As we explore the creation of an audio manager with some built-in functions, we will learn how to play different sounds, depending on which type of object is picked up. We will also learn how to transition to a different *snapshot* depending on which part of the level we are in.

Before we begin to build an audio manager, it is very important to decide what functionality it needs to have. This pre-planning is absolutely critical to good programming in general: prior to writing a single line of code, you should have thought through exactly what you are trying to achieve. In this case, to keep things simple, let us decide on the following features. Our audio manager should:

- hold and keep track of our sounds all in one place so we can see what sounds are being used in the game;
- play the sounds (this means it will need at least one virtual speaker, essentially, an audio source, to play from);

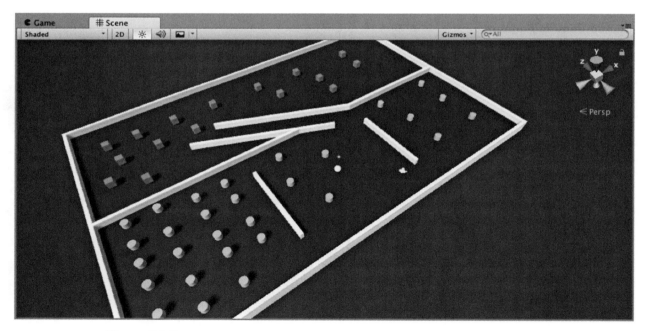

Figure 14.5.
A more complex *Simple Pickup Game*.

- play a random sound from a folder of files (for example, if there is a folder full of 15 gravel footsteps, the audio manager should be able to select a random file from that folder and play it);
- randomize the pitch and/or volume of our sounds;
- handle mixer snapshot transitions so that we can smoothly transition to different mixer group volumes (we will learn what these are later in this chapter).

Creating the `AudioManager` Game Object

Now that we know our programming goals, we can begin the creation of the `AudioManager` functionality. First, right-click within the hierarchy and select "create empty" in order to create a new empty game object. Select this new object in the hierarchy, go to its settings now visible in the Inspector, and attach a new script called `AudioManager`.

SINGLETONS: ENSURING ONLY ONE AUDIOMANAGER AT A TIME

Before writing the actual code for our audio, we need to understand the concept of a *singleton*. When you create a global object such as an audio manager, one that will contain functions used in various scripts, it is essential that there is only one of this type of object in your game. This is particularly important when the game reloads a scene. For example, if the player dies, many of the assets in the game are reset to an earlier point in the timeline. When this changeover occurs, a second instance of the audio manager should not appear. If it did, not only might it cause a variety of internal issues, but—at best—we would hear each sound played twice. The audio clips would be loaded into the game multiple times, and functions would be called unnecessarily. The way to ensure that a singleton remains the only working copy in a project is by adding the following code in the `Awake()` function:

```
void Awake () {
  if (instance == null) {
    instance = this;
    DontDestroyOnLoad (gameObject);
  } else if (instance!= this) {
    Destroy (gameObject);
  }
}
```

The `Awake()` function is another built-in *Unity* function. `Awake()` runs even before `Start()` runs. It is useful for placing events that must occur before the game starts.

Remember, the `AudioManager` script is attached to an object that has temporarily come into existence in the game. However, in the script's `Awake()` function, before the engine can perform any tasks, the game must check if an instance of the `AudioManager` object already exists; if so, this newer one will be deleted. Thus, with these few lines of code, a singleton's "solo-ness" can be preserved.

Implementing the Score

The first line of the *if* statement in `Awake()` checks whether an instance of the `AudioManager` already exists. If `instance == null`, this means that there is not currently another `AudioManager` in existence, so the engine continues along this path, exiting the *if* statement, and moving onto `DontDestroyOnLoad()`. `DontDestroyOnLoad()` is another built-in *Unity* function which accepts a `gameObject` and prevents it from being destroyed when a new scene is loaded. This allows `AudioManager` to continue existing even after the player dies and a scene is reloaded.

The `else if` section of the code is reached when an `AudioManager` does already exist in the game.[7] In this case, the `Destroy()` function of *Unity* destroys the `gameObject` the `AudioManager` is attached to, ending its existence and allowing the already existing one to survive.

DECLARING VARIABLES

We will need to store some audio clips in `AudioManager`. To do so, let us first declare variables with which to hold these audio clips:

```
public AudioClip pickupClip1;
public AudioClip pickupClip2;
```

Now, in the Inspector, let's drag our audio files from the project window into their corresponding `pickupClip1` and `pickupClip2` slots so that these variables become associated with the proper files.

We will also need virtual speakers to play the sounds. Let us add variables to hold two audio sources:

```
public AudioSource SFXSource;
public AudioSource musicSource;
```

Then, once again in the Inspector, create two audio sources and drag one into each variable so that each of these variables is associated with a virtual speaker in the game.

WRITING THE FIRST AUDIO FUNCTION

We will now create a simple function that plays a sound at a specific point. Again, because we are creating this function in our `AudioManager`, we will not have to rewrite the function in each script when we wish to call it. Rather, we will simply call upon it from the `AudioManager`.

In order for functions to be useful in other scripts, they require inputs. As we previously discussed, a function requires both input parameters and instructions on what to do with them. Since we want to be able to tell the function which sound to play, we will add a clip input parameter. We will also add an input parameter to determine which audio source we will use to play each sound. Then, in the body of the function, we place the instruction to play the audio clip. `void PlaySound()` now looks like this:

```
void PlaySound (AudioClip clip, AudioSource source) {
  source.clip = clip;
  source.Play();
}
```

The input receives an `AudioClip` that is called `clip`, as well as an `AudioSource` that is called `source`. The first instruction `source.clip = clip;` tells the audio source which `clip` to play:

1. `source.clip` is the `clip` held by the variable `AudioSource source`.
2. `=` sets `source.clip` equal to the new `AudioClip`, `clip`, which was input into the function as we called it.
3. `source.Play();` tells the AudioSource to play the clip.

RANDOMIZATION

One excellent way of giving repetitive sounds some variety is by randomizing their pitch and volume so that they vary a small amount each time they are played. Let us explore how to create a pitch and volume randomization function for cases in which more sonic variety would be desired. This function will take the input variables `minPitch`, `maxPitch`, `minVol`, and `maxVol`. Since pitch and volume are both float values, these inputs will be floats.

Unity has a very useful built-in mathematical function called `Random.Range()`. `Random.Range()` accepts a minimum value and a maximum value to define a range. Then, it outputs a random number somewhere within that range. Feeding the input parameters of `Random.Range()` with our "min" and "max" variables, respectively, will therefore result in a random number within that range; this process will be used to randomize our pitch and volume.

First, we will set pitch and volume to a random number in the range using the inputs of `Random.Range()`:

```
volume = Random.Range (minVolume, maxVolume);
pitch = Random.Range (minPitch, maxPitch);
```

Now that we have done this, the variables that store "pitch" and "volume" are now each storing a random value within the provided ranges. However, these settings will still need to be applied to the audio source; the following lines of code do just that:

```
source.pitch = pitch;
source.volume = volume;
```

USING MULTIPLE AUDIO SOURCES

Now, as we play the game so far, we will encounter a problem: audio sources can only play one sound at a time. Therefore, if we were to collect two pickup objects quickly, one right after the other, the second one's sound effect would cut off the first. This behavior creates a jarring audio experience—in other words, it is not immersive, but seems very amateur. Remember, an audio source is like a speaker, and that speaker can play only one musical track at a time—that is, to play a second track, the previous one has to first stop. To solve this unfortunate performance problem, we will need to use a method that allows each sound to play from a different audio source so that multiple tracks can sound simultaneously.

The method we will use is to create an *audio source prefab*. A *prefab* is an object, predesigned to have certain characteristics, that can be used over and over throughout the game. In *Unity*, this is created by building an object in a game scene and then dragging it into the project window, thereby turning it into a file that can be called upon as needed. Creating prefabs allows developers to skip the step of having to build an object from the ground up each time a copy is needed. Each time a collision with a Pickup Object is triggered:

1. An audio source will be generated at that location.
2. The sound effect will play through that audio source.
3. The audio source will be destroyed.

First, though, we will need to create a variable that will hold a reference to the audio source prefab file so that we can tell the script to create an object with an audio source on it each time we need one. We can do this by using the following line of code:

```
public GameObject audioSourcePrefab;
```

Next, we will need to create an audio source game object within the Hierarchy and drag it into a folder in the Projects window to turn it into a prefab. Essentially what happens here is that by dragging the object into our computer's file system, we tell Unity that it is no longer simply a game object in the game; rather, it is now a file on the computer that we can reuse. Next, we drag that file from the Projects window into the public slot `audioSourcePrefab` in the Inspector, thereby associating the prefab with the `GameObject` container. Finally, we will use this line to create an `AudioSource` in the game:

```
AudioSource myAudioSource = Instantiate(audioSourcePrefab.GetComponent<AudioSource>());
```

To Instantiate an object means to create an instance of that object in the game world using the built-in `Instantiate()` function. In this case, we are *instantiating* an `AudioSource` called `myAudioSource`. We tell our `GetComponent()` function to look at what is currently attached to our prefab and find the audio source. So, `GetComponent()` finds the audio source and then creates one in the game world (i.e., creates the speaker we will need to play our sound).

Figure 14.6 shows what the final multi-audio source function will look like. In line 27, it is given the name `PlaySoundComplex`. Lines 27–34 show the various inputs `PlaySoundComplex` will accept. Then, in line 36, `audioSourcePrefab` is instantiated. After randomizing the volume and pitch through `Random.Range`, the sound is played on line 42. Finally, line 44 instructs the function to destroy the audio source after `myAudioSource.clip.length`; it will destroy the `myAudioSource` that had just been created and used, once the exact amount of time as the length of the clip has transpired. In other words, it will destroy itself the instant the audio clip finishes playing.

```
27      public void PlaySoundComplex(AudioClip clip,
28                                   GameObject objectToPlayOn,
29                                   float pitch,
30                                   float volume,
31                                   float minPitch = 1,
32                                   float maxPitch = 1,
33                                   float minVolume = 1,
34                                   float maxVolume = 1)
35      {
36          AudioSource myAudioSource = Instantiate(audioSourcePrefab.GetComponent<AudioSource>());
37              volume = Random.Range (minVolume, maxVolume);
38              pitch = Random.Range (minPitch, minVolume);
39              myAudioSource.pitch = pitch;
40              myAudioSource.volume = volume;
41              myAudioSource.clip = clip;
42              myAudioSource.Play();
43
44              Destroy(myAudioSource.gameObject, myAudioSource.clip.length);
45
46      }
```

Figure 14.6.
The multi-audio source function PlaySoundComplex.

Using Other Scripts' AudioManager

For `AudioManager` to be useful, its functions need to be accessible no matter what script is calling them. Otherwise, its usefulness would be limited to the single, narrow use-case of its "home script," and as a singleton, `AudioManager` cannot exist in more than one instantiation at a time. Thus, the `AudioManager` object must be housed in one script, and other scripts will reference it, which will allow any script to call upon any functions or variables within the `AudioManager`.

At the beginning of any script in which the functions or variables of `AudioManager` must be referenced, a public variable must be declared to hold that reference to the `AudioManager` object:

`public AudioManager AM;`

In the `Awake()` function, the script will be instructed to locate the `AudioManager` instance and set it to the variable container:

`AM = GameObject.Find("AudioManager").GetComponent<AudioManager>();`

`GameObject.Find()` is a *Unity* function that tracks down an object with the name between the parentheses. Adding `.GetComponent<AudioManager>();` tells the script to find the `AudioManager` component (in this case, the `AudioManager` script) attached to the game object. Now, whenever the script sees AM, it will have a direct reference to the game's `AudioManager` instance and will be able to use its functions and assets. If, for example, we then would like to use the `PlaySound()` function in this script, we could write:

`AM.PlaySound();`

Conclusion

Programming may seem difficult, but as you have seen, even this single book chapter can introduce you to a wealth of integral concepts and even get you started with some intermediate applications. While we have focused on C#—the programming language most commonly used in *Unity*—the concepts you have learned will greatly assist you as you delve deeper into programming, whether it be C# or other programming languages. The fundamentals we have discussed (class structure, functions, variables, association, game objects, etc.) will be quite useful in understanding the workings of any game engine. As with all skills, there is no better way to improve programming ability than to practice it with real applications. You are therefore encouraged to move forward and venture further into game programming.[8]

Exercises

1. Search for *Unity 3D's Roll-a-ball Tutorial*. Follow along with the tutorial and create your own version of the game.
2. Write a music loop that will be background music for your game.
3. Create a short musical sound effect for when the ball collides with a "pickup" object.
4. Create a new game object and name it `AudioManager`.
5. Attach a new `AudioManager` script to this game object.
6. In the `Awake()` function, turn the `AudioManager` into a singleton.
7. In your `AudioManager`, create variables to hold your music and sound effects clips.
8. Create your `PlaySound()` function and give it inputs for an audio clip, pitch, and volume.
9. Create volume and pitch randomization in the function.
10. Create another script and create a reference to your `AudioManager`. Use this reference to access your `PlaySound()` function from the other script.
11. Code it so that when the player object collides with a pickup object, your pickup sound effect plays.
12. Play the game and make sure you are hearing sound effects where they should be.

Notes

1. A *script* is a document written in a scripting language. It generally contains all of the code pertaining to a specific game object or type of object.
2. Note that it is written with a pound-sign (#) and not a sharp (♯), though.
3. Remember, *pseudocode*, which we discussed briefly in Chapter 13, is an informal high-level description of the operating principle of "real code."
4. As we learned in Chapter 13, a variable is a piece of memory that serves as a container for a specific type of value.

5. Among other features, variables, functions, and other types are color coded differently, and indentations are automatic.
6. It is standard to declare any variable that will be used throughout the script at the top of that script.
7. The != notation, such as found in instance != *this,* means "not equal."
8. If you plan to work in *Unity*, Unity Technologies has excellent free tutorials on their website.

15

Using Audio Middleware

What Is Audio Middleware?

While game engines are impressive in the tools they may provide for development, they lack many features that are important for the workflow of composers and sound designers. Often, a composer must spend several hours accomplishing an otherwise-simple task in *Unity* that could be achieved quite easily using other methods. As we learned in our previous chapter on *Unity*, creating an object to handle the randomization of sounds for *Where Shadows Slumber* (2018)[1] took not only a great deal of time and effort, but also the assistance of a skilled programmer. Realistically, creating a global musical clock and programming the music to transition properly (for example, when horizontally resequencing a score) is even more complicated and time-consuming, even for skilled audio programmers.

To overcome these obstacles, composers use middleware software to easily create dynamic music and sound structures without having to resort to any complex programming. It is meant to bridge the gap between game engines and audio softwares, providing the composer with a visual interface and functionality that resembles a DAW. The core difference between middleware software and DAWs is that middleware is built to handle *game calls*—messages or instructions sent from the game engine to the audio middleware software—enabling the connection between the game and the audio that allows the music and sound to change in real time according to the on-screen events.

In this chapter we will learn about audio middleware and its functionality, discussing the practical reasons why it is ideal to implement middleware to create game audio whenever possible. We will explore three different industry-leading middlewares—*FMOD, Wwise,* and *Elias*—and how they function. Each of these has its own advantages and disadvantages in terms of application. We will see how each of these middleware options looks and functions. So, by the end of this chapter, you should have the necessary information to decide which software might be best suited to each game project you might encounter.

The Core Advantages of Middleware

One central problem that affects coding audio (and particularly music) in a game engine like *Unity* is the fact that like most engines, it runs in *game time. Game time* is not like the normal, linear concept of time; rather, it is dependent on frame rate. Thus, while the game might generally run with one game-time second corresponding more or less to one real time second, this relationship cannot be relied upon. If the frame rate of the game drops, for instance, one game-time second might suddenly become two real time seconds. Since audio is generally dependent on real time processes, this could cause a complete desynchronization between the music and the game. This tenuous relationship creates many problems for composers in their attempts to synchronize music not only with the game, but even between different musical elements. Therefore, implementing advanced music sequencing techniques directly within the game engine, depending on the needs of the game, can become quite cumbersome and unrealistic.

For instance, say that the composer wants to create a simple musical transition in which a cue begins directly on the downbeat. This task sounds simple enough; however, since *Unity* does not have a real time clock, the composer would most likely have to program the game to wait a certain amount of time and then begin playing the next cue. Herein lies the issue: if the frame rate changes even minutely, the player will experience noticeable desynchronization between the first cue and the second. After all, a difference in fractions of a second could create a perceptible error in the musical alignment. This would be even more apparent when attempting vertical layering: different layers might enter completely unaligned rhythmically.

While it is certainly possible to write scripts within *Unity* to overcome these issues, it takes a great deal of programming expertise to accomplish this. And, even with such skill, to do so requires many hours of programming and debugging. Middleware allows all music implementers to forgo this tedious process by giving them the necessary tools for creating audio functionality.

In summary, middleware is essentially a DAW for game audio. It presents composers with a relatively familiar digital UI that they can use to import audio clips and arrange them however they need for their game. Unlike a DAW, however, middleware is built not for editing audio, but for creating interactive audio for communication with a game engine.

The Big Three: *FMOD, Wwise,* and *Elias*

FMOD Studio

FMOD Studio[2], by Firelight Technologies, is one of the industry's leading middleware softwares, serving as the audio solution used for many well-known games, including *Dark Souls III* (2016),[3] *Dead Rising 4* (2016),[4] *Celeste* (2018),[5] *Flower* (2009),[6] *Kingdom Come: Deliverance* (2018),[7] *Just Cause 3* (2015),[8] *Just Cause 4* (2018),[9] and *Minecraft* (2011).[10] Of the three middlewares we will discuss in this chapter, its visual interface most closely resembles that of the average DAW. Thus, it is an excellent starting point for understanding how middleware software works and why middleware may be an excellent choice for implementing audio and music in games.

FMOD almost completely bypasses the coding necessities required by interactive music, giving composers the tools they need to create such functionalities within its interface. In *FMOD*, it is fairly trivial to set up a complex musical structure. While programming audio functionality from scratch can require complex code, using *FMOD* to

Using Audio Middleware

induce a musical change is often as simple as giving *FMOD* a command to "play music now." Often, this step involves inserting a single, simple line of code in one of the game's scripts, instructing *FMOD* to begin playing a complex musical structure that has all been programmed through *FMOD*'s UI, with perfect synchronization and transitions, completely avoiding any of the potential timing issues that might have occurred natively in *Unity*.

FMOD's close visual resemblance to a DAW sets it apart from other middlewares like *Wwise* and *Elias*. Just like a DAW, it utilizes a horizontal timeline view of musical tracks and sound effects, with different tracks stacked on top of each other if their audio overlaps in time. At the top of the timeline view, a right click allows the creation of different markers (including tempo markers, transition points, looping regions, time signature changes, and more), just like in leading DAWs such as *Pro Tools*, *Ableton Live*, and *Logic Pro X*.

One very helpful feature of *FMOD* is that multiple tempo and time signature markers can be placed for each cue. Loops are simply set up within an event's timeline by surrounding the audio to loop with a "loop region," as shown in Figure 15.1. Transitions are set by right-clicking in the timeline and creating a transition marker, as denoted by the green arrow in Figure 15.1.

In *FMOD*, there is also easy access to different means of controlling *parameters*. Parameters are values that are checked in every frame, thereby allowing dynamic changes in real time. Parameters are extremely useful because they provide a way to attach a particular game call to a matching musical event. For example, the volume of an intense drum track could be attached to the player's proximity to an enemy, causing the track to gradually increase in volume as the enemy grows closer and the danger increases.

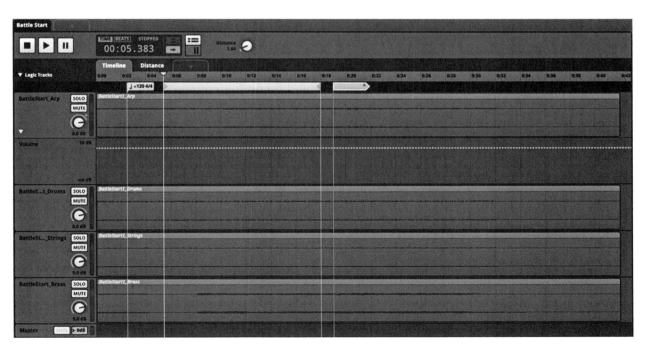

Figure 15.1.
A loop region in *FMOD*. FMOD Studio (2.0), Firelight Technologies (2019).

Implementing the Score

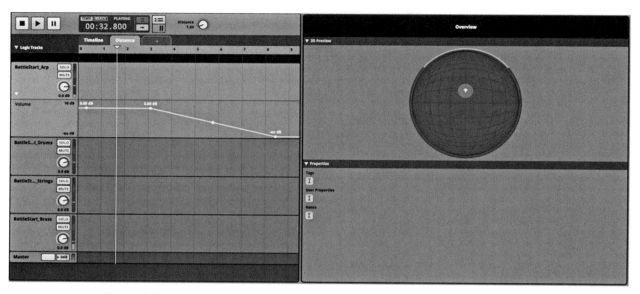

Figure 15.2.
An automation line controlling the "distance" parameter in FMOD.

Clicking the "plus" button next to the "Timeline" tab creates a new tab for a parameter. The small knob above that tab may be used to set the parameter's initial value. Interestingly, a parameter tab also shows your musical tracks, just like the timeline tab does. Much like in many DAWs, right-clicking on different features (for example, the volume knob) allows the creation of an automation line (Figure 15.2).

The "effect deck" holds all of the audio DSP effects attached to a specific track. It appears at the bottom of the screen and greatly resembles a track effects chain in, say, *Ableton Live*. A number of built-in effects can be added within this section. A very user-friendly and useful feature of *FMOD* is that any of the effects knobs can be automated based on any input parameter being sent from the game. For instance, a high-pass filter effect could be added to a music track, and the filter cutoff frequency could be easily automated to move up and down, depending on how fast the player is moving. Figure 15.2 demonstrates the automation line of the "Distance" parameter, while 15.3 shows the window for adding a parameter to the project. "Distance" allows the automation of different factors based on the player's distance from a certain object in a game. The left side shows the automation line. As the white cursor is moved to the left or right, the distance changes on the right side 3D Position view, the volume increasing or decreasing as the proximity to the object changes.

FMOD.io is a feature built into *FMOD* that provides users with a large soundbank of sound effects that are properly formatted for *FMOD* so that they can be easily and quickly dragged into any *FMOD* project. FMOD.io also features an online store where thousands more sound effects can be found.

FMOD uses "on events" to connect game calls with audio actions. Within each event is a certain number of tracks in which can be placed music and/or sound effects. The visual interface is laid out quite logically. In this example *FMOD* screenshot (Figure 15.4), the Events tab on the left contains a list of all of the events. Clicking on an event in the list brings up all of the audio tracks that are part of that event. In the main part of

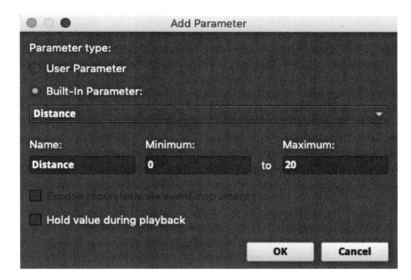

Figure 15.3.
The "Add Parameter" window in *FMOD*.

Figure 15.4.
The Events tab in *FMOD*.

the *FMOD* window, the timeline view of the tracks is similar to that of a DAW, allowing edits and fades to be made directly from the UI.

"Triggers" may be used to channel quick responses for short events, like stingers. These can be placed within the event's timeline. While not shown in the figures, bear in mind that there is also a "Trigger Behavior" panel available in *FMOD*. Expanding this panel shows a number of tools for determining such things as when these stingers will play, how often they will play, and at what point in the timeline they will play.

All in all, *FMOD* is an excellent middleware software with an easily understandable and familiar UI. Because of its resemblance in both look and function to most DAWs, it is a good place to start when first exploring the use of middleware. However, there are other fantastic softwares, each with their own unique takes on the audio implementation process, which offer other features that are also worth exploring.

Thus, we will now take a look at another commonly used middleware software: *Wwise*.

Wwise

Wwise[11] is a complex, yet fairly intuitive middleware software developed by Audiokinetic, which now (as of May 2020) is owned by Sony. Because of its all-encompassing functionality and easy integration with game engines, it is among the most prominent middleware softwares in use today. Audiokinetic offers a great deal of excellent tutorials on its website (audiokinetic.com); however, this chapter will focus on music integration, giving enough information to make a confident start, as well as demonstrating some fascinating uses of the software that go beyond simple implementation.

Like *FMOD*, *Wwise* makes the implementation process of musical loops and complex game-influenced forms easy. This is because *Wwise* has a real time clock and parameters that can be set by the user to determine the relationship of each musical segment to another. For example, upon first importing a segment, the start and end points of a loop can be set. Any audio that exists before the start point can be used as "pre-entry," a segment that serves as an intro into the loop. Using pre-entry can greatly smooth a transition because this introductory segment can be played on top of the previous track as it is fading in.

It is also possible to set the beats per minute (bpm) of each segment, which is crucial for creating transitions that occur at the end of a bar, beat, or phrase. However, the pre-entry might actually start fading in over the first segment three beats before that segment has finished playing. Likewise, another parameter called the "post-exit" exists, serving a similar purpose as the pre-entry; however, it instead plays after the first segment has already transitioned to the following segment.

Wwise is also useful in situations in which the music should transition between two different tempos. There are a variety of transition settings built into *Wwise* that can be used to easily mold the transition as the audio programmer sees fit.

Using Wwise to Create Dynamic Music

Most of *Wwise*'s musical functionality can be edited in the "Interactive Music Hierarchy," a section of the UI containing a variety of modules dedicated to creating musical structures (Figure 15.5). The interactive music hierarchy provides an excellent interface for creating musical transitions based on state changes—we will discuss this functionality in further detail later in this chapter.

A large part of *Wwise*'s functionality is based on the concept of *containers,* folders that hold a variety of files sorted in a specific way. Each container type works differently and serves a different purpose. One commonly used type is the "random container"; if audio files are placed in a random container, a random audio file from that container will play

Using Audio Middleware

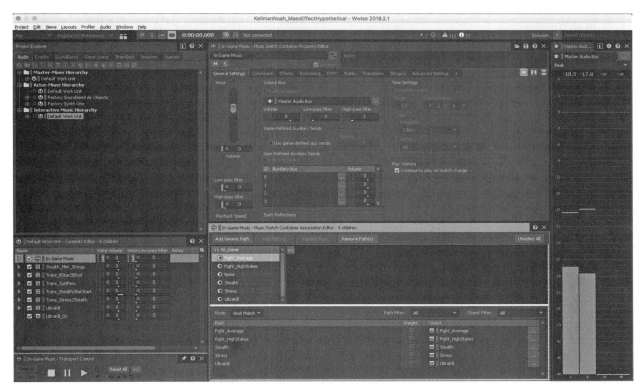

Figure 15.5.
The Interactive Music Hierarchy in *Wwise*. *Wwise* (2019.1.6.7110), Audiokinetic (2019).

each time the container is triggered to play. This functionality is commonly used not only for the playing of random musical segments, but also for sound design. For example, 15 different versions of a "grass footstep" could be placed into a random container. Then, each time a grass footstep needed to be triggered, one of those 15 different files would be chosen at random by *Wwise*.

An incredibly useful container, particularly for the horizontal resequencing of music, is the "music playlist container." This can be used to set up the order in which music tracks will play. Tracks within a music playlist are then further placed into groups. The groups have different functions:

- *Sequence Continuous*: Tracks in this group will simply play in sequential order, one after the other. This is the group type to use for designing a linear musical form.
- *Random Continuous*: Tracks in this group will play in a randomized order. It is even possible to set how likely a given track will be chosen by setting its "weight." A higher weight means it has a higher probability of being chosen, and a lower weight means it has a lower probability of being chosen. The "loop count" determines how many times a segment will loop—and, as with many parameters within *Wwise*, it can also be randomized. Furthermore, if the entire playlist should repeat instead of stopping once all the tracks have played through once, the playlist itself can be set to loop as well. When the entire playlist loops, the same sequence of logic will be carried out again. If, for example, your playlist contains a random continuous group, these tracks will play again in a randomized order.
- *Sequence Step*: In this group type, multiple musical segments are placed. On the first play-through of the sequence, the first segment will play. Then, on the repetition, this

group type will automatically "step" over the first segment and go directly to the second. This continues on, sequentially.

Using music segments and groups, a complex musical hierarchy can be created. For example, there might be a larger Sequence Continuous group called "Combat Music," which is organized as follows:

Combat Music (Sequence Continuous):

1. Music Segment: Combat Musical Intro
2. Random Continuous Container: Random Combat Music
3. Sequence Step Container: Combat Interlude
4. Music Segment: Combat Music Outro.

This hierarchy would function by first playing the Intro. Then, a random track from the "Random Combat Music" group would play, followed by the first interlude segment from the "Random Combat Interlude" group, finally ending with the Combat Music Outro. Then, if we set this group to loop, this entire sequence would repeat; however, on the second repetition, the "Combat Interlude" group would skip Interlude 1 and play Interlude 2 instead. By experimenting with this hierarchical structuring, complex, highly-dynamic musical forms can be created or generated.

States and Switch Groups

During our discussion of state-driven versus player-driven music systems in Chapter 8, we spoke about how music develops as the game state changes. *Wwise* makes setting up such musical changes very easy and intuitive by using "states" and "switches." *States* in *Wwise* are used for global changes in the gameplay. For example, when the game switches between an exploration state and a combat state, a *Wwise* state should be used to transition between the appropriate musical tracks. This can be a particularly effective technique because *Wwise* states allow the option of controlling the length of the transition time. *Switches*, on the other hand, are used for immediate, quick changes that need to occur, like switching between different types of footstep sounds. Here, no global change has occurred in the gameplay, but a local change—which sound container is being used—is necessary.

Changing between musical states works best when the music is placed in a "Music Switch Container": before importing music, go to the Project Explorer under the Audio Tab and find the Default Work Unit under the Interactive Music Hierarchy. Right-click on the Default Work Unit and create a new "Music Switch Container." Importing the audio files into this container allows us to move on to the next step: using "switch groups" and "state switches" to set up dynamic musical reactivity.

States are created under the "Game Syncs" tab in *Wwise*. Under the "Default Work Unit" in the Project Explorer, create a state group by right-clicking, selecting "new child," then selecting "state group." Individual states can then be created by right-clicking on the state group itself and following the same process. Finally, associate each state with the desired music track by using the music playlist editor.

In order to switch between states, enable "State" as the switch type for the music switch container: click on the music switch container, then go to the general settings tab and select "State" under the "Switch Type." At the bottom of the screen is the "Music Switch Container Association Editor," where individual musical elements can be dragged in order to associate them with the proper state. Once this process is complete, enter the "Transitions" tab of the music switch container, a powerful tab that gives control over how the transitions between states occur. Transitions can be set to occur immediately, at the next bar of music, or on the next beat, all with or without transitional segments, and even more options exist beyond these. Depending on which states are transitioning to each other, exactly which type of transition should occur can be chosen. Conveniently, *Wwise*'s built-in playback allows easy testing of all of these different possible state transitions.

Wwise and Parameters

One other important feature of *Wwise* worth our attention is the "Real-Time Parameter Control," or RTPC. RTPCs can be attached to musical cues and allow game data to determine how parameters should be affected within *Wwise*.

Consider the following example: the programmer sets the volume of a drum segment to be correlated with the distance the player is from an enemy. As the player approaches this enemy, the volume increases, alerting the player that he is walking into danger. As the player backs away from this enemy, the volume decreases, signaling that he is safer.

Our companion website example demonstrates the use of state switches and RTPCs through a hypothetical score to *Mass Effect: Andromeda* (2017).[12] ▶(15.1)

The Power of Sending Game Calls Back to the Game Engine

In most traditional games, the game engine controls the audio system by sending game calls that inform the audio system when a sound or music track should change, or play. However, when gameplay is audio-driven, the audio engine takes the lead, sending game calls back to the game engine and having its own effect on the gameplay. In such systems, the gameplay itself may be directly connected to the music, ensuring that in-game events occur in sync with the musical score. Audio-driven gameplay is difficult to code, but middleware makes it feasible by providing an accessible infrastructure for composers and programmers alike to use.

For his score to *The Worst Grim Reaper* (2020),[13] by studio Moon Moon Moon, composer Mark Benis devised a system that uses cues from *Wwise* to trigger in-game events with the score. Lyrical content is paramount to the aesthetic of the score of *The Worst Grim Reaper*, which features a singing narrator (the voice of developer and musician Mark Lohmann). The team's objective was to create a gameplay experience in which the player can, in essence, play a central role in creating the lyrical structure of a song, while also creating a game world that directly responds to the most meaningful words in the song's lyrics. Thus, the player triggers new lyrics by completing specific actions, and also sees those lyrics changing the game world as well. Designing such a system was no easy task; despite the unpredictable formal structure of the songs, the relationship between the player, the lyrics, and the game world must maintain coherence while simultaneously being reactive. Creating an audio-driven system in *Wwise* allowed Benis and Moon Moon Moon to do just that.

Having the lyrics influence the game engine's actions, rather than the opposite, was an integral part of achieving the right emotional experience for the player. As Benis said,

> The emotion we wanted was for the players to feel that they're creating their own song by triggering lyrics and having the story told through a "singing narrator." For us, synchronization is so important because if you think about it the other way—something appears on screen and then lyrics play—that's a purely reactive and a bit more lifeless system. By having the lyrics as game triggers, they proactively work to create the experience and I think that makes a more immersive experience.[14]

Using *Wwise*, he designed a system in which the game engine and audio engine are constantly communicating. Game calls are sent from the game engine to *Wwise*, inducing musical changes; however, more game calls are then sent back from *Wwise* to communicate information back to the game engine. In Benis's words, "To accomplish this granular level of sync, we developed a custom cue with three different types: lyric, fade and chord. Lyric cues would tell the game engine when to show a certain string of lyrics, fade cues would indicate what game objects to animate appearing, and chord cues would specify what the current chord of the song is (to tune other sound effects to that chord)."[15]

In other words, when the music reaches a specific lyric (a lyric cue), a call is then sent back from *Wwise* to the game engine, triggering an in-game animation. Additionally, when the music changes its chord (a chord cue), a game call is sent back to the engine telling it the new tonality. Since the game engine is constantly informed about the current musical key, it can then tell *Wwise* in which key a musical sound effect should be played when it is triggered. At the technical level, this was accomplished by creating a different state in *Wwise* for each chord; as the player triggers a musical sound effect, the engine uses its knowledge of the musical key to ensure that *Wwise* is set to the right key state, thereby ensuring musical cohesion. ▶(15.2)

Playdead Studios' Creative Uses of Wwise for Inside

With an audio team lead by composer and audio designer Martin Stig Andersen, Playdead Studios created *Limbo* (2010),[16] which has an engine-driven game system. Each time the player character dies, the music stops and reloads from the beginning. While this series of events is quite common in the course of gameplay, Andersen realized the negative effect that this death-respawn sequence of events could have on immersion because of the break in the audio flow. Thus, when he and the Playdead programming team embarked upon their next game, *Inside* (2016),[17] they decided to try the concept of audio-driven gameplay, exploring *Wwise*'s capabilities to craft a solution: an audio-driven game system.

What the team determined was that if they were to make it so that the music would be continuous throughout the scene, even through the death and respawn, they would have to "bake" the rhythm into an audio loop; then, at the respawn, the timing of game animations would have to adapt to the audio, rather than vice versa. The game therefore has to constantly be measuring the position in the audio at the time of the respawn event, and then orient itself based on that information.

Using Audio Middleware

The most difficult part of executing this type of scenario is that the game must operate in linear time rather than game time. A change in frame rate could easily throw off the timing of a puzzle, creating a problem for the player, who could not possibly determine the solution when the timing is off. Therefore, the game must be responding to the music itself, which exists in real time. According to Andersen, the solution to this problem was to allow the audio engine to run continuously, thereby creating a system where the audio functionality would always remain unaffected by player death (unless otherwise specified).

Prior to building the *Wwise* file for *Inside*, it was decided that the sound would be split into two major categories: those that restarted upon respawn (for example, any sounds attached to actions that would be carried out again after the player respawned) and those that continuously played even after death (for example, music and ambience, which simply exists whether the player is alive or not).

In the following example, seen in Figure 15.6, the player must move in perfect rhythm with the score, turning the player character's body perfectly with the NPCs on the screen, who also turn in time with the music. The music's structure comprises a looping four-bar figure. In this scene, during the first half of the loop—the first two bars—the player character (a little boy) has to march. During the second half of the loop, he has to stop and stand still. If he were to do the opposite (e.g., stand still in the first half of the loop), he would be killed. Rather than the game engine being in control, it is the music that ultimately determines when the other characters will be walking.

In this case, it was important to ensure that depending on when the boy dies, he respawns at the right point in the loop. For example, if he dies during the second half of the loop, there could be a substantial risk that he might actually end up respawning during the first half of the loop, when he is supposed to be moving. If this were to

Figure 15.6.
A rhythmic puzzle in *Inside* (2016).

Implementing the Score

happen, he would instantly be killed again before getting a chance to move. In order to prevent this problem, the Playdead team created a two-bar loop that cues as soon as the screen fades to black. This extra loop ensures that the player will not respawn during the first half of the loop and instantly die. ▶(15.3)

A Foray into Middleware Programming

The use of middleware does, for the most part, relieve the composer of programming duties, while simultaneously making the programmer's job much simpler. However, indie game composers may sometimes find that they still have to—or perhaps even want to—do some basic programming themselves. When it comes to controlling middleware with code, a little coding knowledge can go a long way. The following is a list of the most commonly used script functions that control functionalities in *Wwise* (written here in the programming language we have been using thus far, C#):

- `AkSoundEngine.PostEvent(string eventName, gameObject);`
  - This is a simple *Unity* function to tell *Wwise* to trigger an event. The script is told that it is using a function from `AkSoundEngine` called `PostEvent`.

- `AkSoundEngine.SetState(string stateGroupName, string stateName);`
  - This function is used to change states from a script. For it to work properly, both `stateGroupName` and `stateName` must match their corresponding names in *Wwise*.

- `AkSoundEngine.SetSwitch(string switchGroupName, string switchName, gameObject);`
  - This function is similar to the `SetState()` function, but is used to change switch groups.

- `AkSoundEngine.SetRTPCValue(string RTPCName, float RTPCValue, gameObject);`
  - This function allows an RTPC value to be set from the *Unity* script.

Elias

Wwise is meant for both sound designers and composers alike. Because it covers a broad spectrum of usability, it is a fairly complex, feature-rich software; however, the UI is less approachable than other middleware options. *Elias*[18] is a newer middleware that focuses specifically on music implementation, and it is quite simple to use. Its interface, shown in Figure 15.7, is visually similar to the clip view in the DAW *Ableton Live*. Each column represents a track, while each row is a different "level." Different clips can be dragged into "boxes" that correspond with each track. Each box contains its own audio file. However, additional audio files containing variations of the original can also be added into the box, which will cause *Elias* to randomly choose one of those variations when that box is triggered.

Whereas navigating *Wwise* requires a great deal of switching between different editor windows, also requiring the user to properly associate all of the objects, *Elias* streamlines

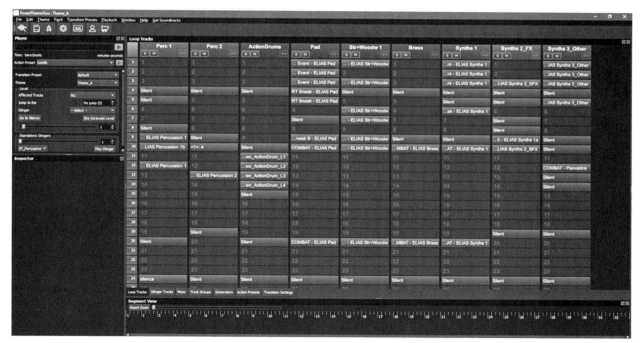

Figure 15.7.
An example of one of Robert Lundgren's themes for *Mutant Year Zero: Road to Eden* (2018) in *Elias*.
Screen captures of *Elias* taken by Robert Lundgren and provided with permission for use in this text.

the process of designing a dynamic music structure, squeezing a great deal of functionality into only a few windows. The UI is set up in vertical "levels." Notice how the first row has a "1" at its left-hand side, indicating its level number. If *Elias* receives a game call indicating a state change, the level of a theme will shift what is currently playing to rows which correspond with that level, providing a functional way to cycle between different types of layering and track complexities based on this level parameter. For example, one might design the score such that the level corresponds to the intensity of the score. At lower levels, the rows might contain low-energy music segments like ambient synths and string pads. As the level is set higher, the rows corresponding to higher numbers might therefore contain more intense musical tracks, like bombastic percussion and loud, high-frequency synths.

Much like musical transitions in *Wwise*, it is possible to specify how the transitions between different levels occur. If a specific track should be silent, one can add a red "track level," which will cause that track to be silent when that box is played. Also like *Wwise, Elias* allows work using MIDI. MIDI files can be added to the layout, just like audio files might be. The "generator," or virtual instrument, that will play that MIDI back can also be easily selected.

"Transition presets" are another useful *Elias* feature that allows the easy reuse of transition settings. "Agility settings" make a default start-stop and transitional setting across all music tracks. *Elias* also includes a "smart transitions" feature, which uses artificial intelligence to analyze musical structures and select the best possible transition points, thereby streamlining the transition creation process.

In the "Stinger Tracks" tab (Figure 15.8), stingers can be added which will function as part of the musical whole, playing at specific musical points as determined by the user.

Implementing the Score

Figure 15.8.
The Stinger Tracks tab for *Mutant Year Zero: Road to Eden* (2018) in *Elias*.

"Custom agility beat points" allow an exact amount of control over where each stinger might play. Stingers that help transition from one theme (or cue) to another can also be easily placed. Different stingers can be set to play only at specific points in a musical cue, thereby ensuring that it is melodically, rhythmically, and harmonically congruent, and that it can transition smoothly to the next track.

Elias Case Study

Mutant Year Zero: Road to Eden (2018)[19] is a strategy game that features real time exploration and stealth modes, as well as a turn-based combat mode.[20] The mysterious post-apocalyptic game setting features a dark, thoughtful underscore by composer Robert Lundgren.[21] As the game itself is based upon a 1980s-era tabletop role-playing adventure game of the same name, Lundgren drew influence from 1980s-era synth-based music. The music system in *Mutant Year Zero: Road to Eden*, created with *Elias*, uses vertical layers of instruments for its musical themes, combined with an intricate horizontal system that develops based on gameplay, allowing a reactive score that changes based on the player's chances of winning a battle, as well as many other elements within the game.

Understanding how Lundgren set up the musical functionality in *Elias* is quite fascinating. Luckily, he graciously took the time to walk the author through the entirety of the music system. Using *Elias*'s level-based UI, Lundgren was able to effectively implement the many different music modes in the game. He used the same setup for each musical theme throughout the game, allowing him to, in essence, design a template to follow when bouncing out music tracks.

In Lundgren's setup (Figure 15.7), Levels 1–3 of the *Elias* theme are reserved for exploration. Level 4 is used for "idle mode," described in further detail a few paragraphs later. For real time exploration, Lundgren used Levels 5–10, transitioning between different levels every 30 seconds to avoid repetitiveness in the score. Level 11 is reserved specifically for when the player is approaching an enemy, choosing whether to attack or simply sneak by. It is, then, somewhat of a stealth cue. You can see that the layer contains only a drum track, which serves as a possible transition into turn-based combat, or simply a short bout of raised intensity.

Levels 12–14 contain the loops for the turn-based combat mode. Note that they have their own "Action Drums" track, which functions slightly differently than the others in that it is directly attached to the player's likelihood to succeed in battle—the less likely to succeed, the more intense the track will become. If the enemy has a large advantage and the player is about to lose the game, the drums will go to their highest level. ▶(15.4)

In order to accomplish these dynamic drum changes, the drum track must be able to travel between different levels without affecting the level of the rest of the music. To account for this, *Elias* provides a "track groups" feature, which allows different groupings of tracks to be at different levels at different times. One could potentially use this feature to have many different instruments operating at different levels simultaneously.

In *Mutant Year Zero: Road to Eden*, musical stingers (Figure 15.8) are used as sound effects. For example, picking up an item that displays a written note causes the sound of a timpani to play over the music. Another use of stingers occurs while approaching enemies: a red circle appears, indicating danger in close proximity, the overall musical intensity increases, and subtle stingers enter over the music in intervals as extra warnings.

When attacking, the music enters the turn-based combat mode. Horn stingers provide musical feedback upon successfully landing a hit on an enemy. The drum track uses four musical levels that increase or decrease in intensity depending on the action. If it is the enemy's turn and the enemy has the advantage (e.g., if the enemy has a strong weapon, which poses an extra threat to the player because of how much damage it can do), the drums will increase in intensity.

Even when the player is idle—not carrying out any specific action and perhaps taking time to think strategically—the music enters an "idle mode," which contains a number of unique "idle mode" stingers. For this reason, Level 4 is critical—since the game is turn-based, the player can actually stand still for quite a while before deciding what the next move will be. At first, Lundgren experienced problems with the music becoming static frequently. If you had been fighting an enemy and drums were at the highest level, for example, they would stay at that level until the player made the next move. To account for this issue, he created "idle" mode. The system enters idle mode after 20 seconds of the player standing still. Idle mode brings the music down to minimum intensity.

Mutant Year Zero: Road to Eden was built using *Unreal Engine*. As the game's system functions, *Unreal* sends a game call to trigger a stinger, then it finds out what bar the music is in and triggers the correct chord. *Unreal* also keeps track of timing such that if a certain window of time passes, it will tell *Elias* to trigger another stinger to avoid monotony. *Elias* knows which stingers will function well with the music because they have been set to play according to what bar the music is in.

Implementing the Score

Lundgren's score is also a good example of how understanding some programming allowed him to actually go far beyond what *Elias*'s built-in functionality could accomplish. He had only a basic understanding of programming languages from working on other games. Additionally, he had not worked with *Unreal* before working on *Mutant Year Zero*; however, he began experimenting with the developers and was part of the game's design process from the beginning, giving him the time and freedom to experiment and learn how to program using *Unreal Engine*.

One advantage of working with *Unreal Engine* is that the engine contains what is essentially its very own visual scripting language, called *blueprints*. Much like Max MSP, using blueprints allows programmers to simply drag and drop objects on the screen, connecting them with inlets and outlets to create functionality and programming instructions without having to write code. Because *Unreal Engine* uses C++ as its core programming language, a fairly difficult language to learn, blueprints are an invaluable asset for team members who need to be able to program but aren't proficient in C++. Lundgren used blueprints (Figure 15.9) to actually extend *Elias*'s functionality, building complex sequences of events that trigger certain functions of *Elias*. This allowed him to trigger different stingers depending on how difficult an enemy level is.

All in all, Lundgren's work on *Mutant Year Zero* is a fantastic example of using many of the skills we have discussed in this book to aid a composer in creating a successful score. Involved early in the game design process, he was able to play a key role in developing a music design for the game, creating a system that would effectively compliment the gameplay. His use of middleware gave him the tools to create the score without having to program to build it; yet, when he wanted to extend its functionality even beyond what *Elias* could accomplish within its interface, his prior knowledge of programming served him well. Lundgren was able to use *Elias*'s visual interface and level system to build a smooth and effective underscore system for *Mutant Year Zero*. In addition to demonstrating *Elias*'s particular usefulness as a middleware, Lundgren's score is an excellent example of the freedom that middleware gives us as composers in general, as well

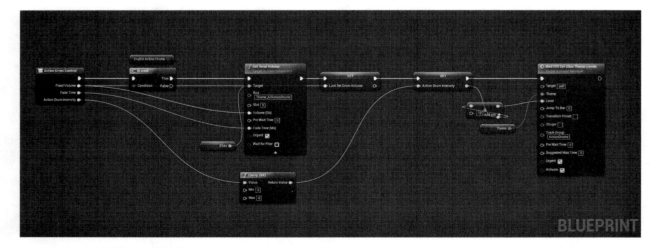

Figure 15.9.
Example of Robert Lundgren's music system blueprints in *Unreal Engine* from his score to *Mutant Year Zero: Road to Eden* (2018).

as the time we can save both ourselves and the developers by being able to create the music system ourselves.

Conclusion

Middleware is an indispensable tool for the modern video game music composer. If video game music is defined by its functionality, middleware allows much of that functionality to be conceived and implemented. Each one brings its own strengths and workflow improvements: *Wwise* provides a detailed, highly-sophisticated environment in which to design a musical soundscape; *FMOD* and *Elias*, on the other hand, offer simpler UIs. *FMOD*, overall, offers an excellent graphical UX thanks to its resemblance to a DAW, and is therefore an excellent "first middleware choice" for the composer new to middleware—though it is by no means amateurish or simplistic. *Elias* is music-centric, making the creation of dynamic musical structures as convenient and quick as possible. In fact, in many cases, *Elias* is used in combination with *Wwise* or *FMOD*—*Elias* for music implementation and *WWise* or *FMOD* for sound design.

Each practical example discussed in this chapter presented the audio team with a unique technical challenge. In a competitive industry with a quickly growing independent market, it is essential for anyone who wishes to be a composer to be prepared with a well-rounded set of skills that extend beyond "mere" music composition. For nearly any game being developed on a limited budget by a small team, a composer who also has experience and knowledge of audio programming and an audio middleware software is a much more attractive hire than a composer who does not. For one, a composer in the former category could begin the writing process with a solid understanding of potential technical "danger zones" which might occur later on, composing in such a way that accounts for these while interacting with game states and game parameters. Overall, rather than requiring constant "hand-holding" by the development team, a composer with such experience can easily become an integral part of the audio-programming process, proposing ideas and personally implementing them as the game develops.

Exercises

1. Open up a new *Unity* session, or download an example session from *Unity*'s website.
2. Choose either *FMOD, Wwise*, or *Elias* and visit the official website for that middleware. Follow an official tutorial about how to integrate the middleware with your *Unity* session.
3. Once you have successfully integrated the software with *Unity*, begin by importing a single music track and placing it into a playable event. Set this track to loop in your middleware.
4. In your *Unity* session, set it up so that when you click play, the music will automatically begin playing.
5. Continue to follow along further with the official middleware tutorials, experimenting with its possibilities. Using what you have learned, implement a state system that

allows you to receive game calls from *Unity* in your middleware session and switch between three different musical states (much like we saw in our *Wwise* example).

6. Continue to experiment until you have fully integrated your game's soundscape with your middleware software of choice.

Notes

1. *Where Shadows Slumber*, mobile, developed by Game Revenant (New York: Game Revenant, 2018).
2. *FMOD* Studio (2.0), Firelight Technologies (2019).
3. *Dark Souls III*, game disc, developed by FromSoftware (Tokyo: Bandai Namco Entertainment, 2016).
4. *Dead Rising 4*, game disc, developed by Capcom Vancouver (Redmond, WA: Microsoft Studios, 2016).
5. *Celeste*, download, developed by Matt Makes Games (Vancouver, BC: Matt Makes Games, 2018).
6. *Flower*, game disc, developed by Thatgamecompany (Tokyo: Sony Computer Entertainment, 2009).
7. *Kingdom Come: Deliverance*, game disc, developed by Warhorse Studios (Hōfen, Austria: Deep Silver, 2018).
8. *Just Cause 3*, game disc, developed by Avalanche Studios (Tokyo: Square Enix, 2015).
9. *Just Cause 4*, game disc, developed by Avalanche Studios (Tokyo: Square Enix, 2018).
10. *Minecraft*, download, developed by Mojang (Stockholm: Mojang AB, 2011).
11. *Wwise* (2019.1.6.7110), Audiokinetic (2019).
12. *Mass Effect: Andromeda*, game disc, developed by BioWare (Redwood City, CA: Electronic Arts, 2017).
13. *The Worst Grim Reaper*, download, developed by Moon Moon Moon (Utrecht: Moon Moon Moon, 2020).
14. Mark Benis (composer of *The Worst Grim Reaper*), interview with author, April 25, 2019.
15. Ibid.
16. *Limbo*, game disc, developed by Playdead (Copenhagen: Playdead ApS, 2010).
17. *Inside*, game disc, developed by Playdead (Copenhagen: Playdead ApS, 2016).
18. Kristofer Eng and Philip Bennefall, *Elias* (Stockholm: Elias Software, 2019).
19. *Mutant Year Zero: Road to Eden*, game disc, developed by The Bearded Ladies Consulting (Oslo: Funcom Oslo AS, 2018).
20. *Turn-based combat* refers to a classic system of gaming combat in which both the player and the enemy take action turns.
21. Robert Lundgren, interview with author, May 16, 2019.

16

Essential Skills for Working with Developers
Version Control and Optimization

Introduction

When working in the video game industry, composers are often required to understand and use methods, workflows, and tools that fall completely outside of the realm of "strict music composition." *Version control,* a method used by development teams to keep track of changes made to a game and to ensure that each team member's asset files are current, is one such example. Version control operates by hosting a master copy of the game software on a server, while team members make changes only to local copies, which are stored on their own computers. Using version control can be complicated, but its use is both essential and ubiquitous. Therefore, in this chapter you will learn the basics of version control workflow, gaining an understanding of the fundamentals of proper use, thereby increasing your value as a development team member.

Another important "non-musical" tool a game composer should have is an understanding of the audio optimization process. When working on a game, a composer will be given a certain "audio budget," determining the maximum limit of computer processing power, memory, and drive space the audio can use once the game is finalized. In many cases, once the audio is composed and created, the audio's "footprint" must then be significantly reduced in order to ensure that it conforms to the specifications of the audio budget, all so that the game ultimately runs smoothly. In this chapter, we will discuss the optimization process, the basics of how and why it is used, and some common audio optimization techniques.

Version Control

Overview

Version control is a system for storing a current copy of an entire game on a server, as well as keeping track of all changes made to the game by any member of the development team at any point in time. It should be clarified that version control is not a software itself, but rather a system for which many different software solutions exist. A "delta backup" system (much like Apple's incremental backup software, *Time*

Machine),[1] version control keeps track of exactly which files in a game are changed each time and by whom, and allows the developer to go back to an earlier version if something goes wrong.

Version control also allows team members to organize their changes and stay up to date without constantly passing the game file back and forth between workstations. Often, a large number of people are working together on the same game concurrently. It can quickly become very difficult to keep everyone's files up to date if each time a change is made, one has to send the entire team a brand new compressed file of the game. In such a case, each team member would have to download and unpack the file, then apply all their new changes to that file, and send yet another copy of the game to the entire team. Obviously, this method would be highly inefficient at best.

Whenever someone makes changes to the game using version control, however, those changes are then applied to the server version, and a detailed log is kept. Therefore, each team member can choose to receive those specific changes without ever having to go through the tedious process of uploading and downloading the game each time it is updated.

Take, for example, a team working on a medieval fighting game: Kellie is working on the "SwingSword" sound effect, while Todd is upstairs working on the animation for swinging the sword. Meanwhile, Jacquelyn is fixing a bug in the code that causes the sword object to get caught mid-swing. Without version control, all three of them might finish at around the same time, sending conflicting versions of the game to the rest of the team. Someone would have to unpack these three versions and put all of the new changes together: a time-consuming and unnecessary process, prone to mistakes and complications.

With version control, however, Jacquelyn finishes her bug fix at 11:12 a.m. and uploads it to the server. Todd sees this change come in and downloads it immediately to his computer. He notices the bug fix and is able to see the old animation in more detail. He therefore does a better job in creating the new animation, which he finishes around 11:25 a.m. Kellie sees Todd's new animation on the server and downloads it. This makes her job easier, because now she can see exactly how long the sound effect needs to be. She finishes at 11:35 a.m. and sends out the version of the game with the sound effect implemented. By 11:36 a.m., everyone on the entire team has a functioning current version of the game with an updated animation and sound effect, and no one had to expend more effort or take more time past the normal extent of their job duties.

While advanced users of version control tend to use *command line*,[2] a simple way for anyone to use version control is by using a software application with a UI specifically built for that purpose. One such application is *Source Tree*,[3] which is used quite commonly. Softwares like *Source Tree* provide a simple UI for carrying out the most commonly used and important version control commands.

The style of version control most commonly used today is *distributed version control*. In a distributed version control system, a master version of the game is stored in the *remote*[4] *repository*.[5] Each team member then downloads the game file to a folder on his or her own computer. These folders are called *local repos*; in a distributed version control system, local repos are perfect working copies of the master so that if anything ever goes wrong, the master can easily be restored from one of the other repos.

The Version Control Process

Changes are made to the game by first *committing* them to your local repo, with a note explaining what they are and why they were made, then *pushing* them to the server. Commits are an excellent method for packaging changes together when the changes are related to the same subject matter or solving the same problem. Similarly, unrelated changes often go in a separate commit, thereby keeping everything organized.

Let us now look at a basic step-by-step outline of the version control process:

1. *Fetch*: This tells the computer to find out if any changes have been made since it last updated its local repo.
2. *Pull*: If changes are present, this tells the computer to download those changes to its local repo, keeping your working version of the game up to date.
3. *Commit*: This tells the computer that you are satisfied with the changes that you have made to your local repo and that you would like to save it to the repo's history. Those changes in effect become primed for upload to the server. You should commit frequently as you make changes to the game; however, keep in mind that these changes will be local-only until you *push* them to the server.
4. *Push*: This tells the computer to upload your local changes to the server copy of the project. While the master repo is not necessarily the only version uploaded to the server, often, you will push these local changes directly to the master repo for the ease of keeping everything organized. You should only push to the master once you are satisfied with the changes that you have made, and you are certain that they have not caused any new problems in your local repo. Once you have committed them locally, pushing allows all the other developers on your team to receive them so that everyone can stay up to date.

More Important Terms and Commands

- *Reverse commit*: What happens if you have already committed your changes to your local repo and then realize you've made a mistake that, in fact, crashes the game? You can resolve this by doing a *reverse commit*, which will take you back to the state your game was in prior to making that commit. This means that any changes you committed will be lost; however, so will the mistakes, and you can begin your process again.
- *Branch*: Branches are the linear timelines on which a game's edits are stored. Most commonly, work on a game happens on the master branch. In other words, if you were to push your commits, you would affect everyone's master branch. However, there are cases in which you might want to work for a while without being connected to the master; perhaps you want to test an out-of-the-box idea that has the potential to "make or break" the game. To test this new feature safely, you could open your own *branch*. In doing this, you could continue to push your changes to the server, thereby keeping track of your work as you move forward; as these changes will be applied to a separate branch from the master branch, you would not risk contaminating the master with potentially unstable edits. In Figure 16.1, the repo begins at the bottom of the tree with "Master C1." After "Master C2," this user branches off into her own working version.

Implementing the Score

Once she is happy with that version, she *merges* it (see the following) back into the master. One important consideration in branch use is that, while working on a separate branch, changes concurrently being made to the master branch by others will not be seen. Thus, if you have been working on a separate branch, you may need to check what has been committed to the master in the interim before *merging*.

- *Merge*: If and when you are satisfied with the changes you have been making to your separate branch, you can *merge* that branch with the master, combining all of your new changes with whatever state the master branch is in. This newly merged version becomes the new working master. However, this can sometimes create the problematic scenario of a *merge conflict*.

- *Merge conflict:* This occurs when the changes made on a branch conflict with changes someone else has made in the master. If this happens to you, it is best to get in touch with the person whose changes conflict with yours in order to resolve the issue. From a technical standpoint, the issue is resolved by going through each conflicting change, which might be as little as a single line of code, and deciding whose version to keep.

However, indie studio Game Revenant's lead programmer, Jack Kelly, points out that "for text files (like source code), more experienced users can go through merge conflicts and resolve them line-by-line (which is often the case), rather than choosing yours or theirs."[6]

Case Study: Using GitHub Desktop for *Where Shadows Slumber*

While working on *Where Shadows Slumber* (2018),[7] the developers at Game Revenant—including the audio team PHÖZ—used *Git*, a commonly used distributed version control system, to host the game. Each time one of us made a change, we pushed it to the server. We used the *GitHub Desktop*[8] application (Figures 16.2 and 16.3) to manage our version control system, a basic, effective software solution for managing a git repository. Similar to *Source Tree*, *GitHub Desktop* offers a simple, functional UI.

Let us now walk through a hypothetical scenario.

Taking a closer look at the UI of the *GitHub Desktop*, in Figure 16.2 we can see the "History" tab, which provides a log of all of the commits in chronological order. When one commit is selected, the UI displays any notes attached to that commit, as well as which files were changed. In Figure 16.3, we can see

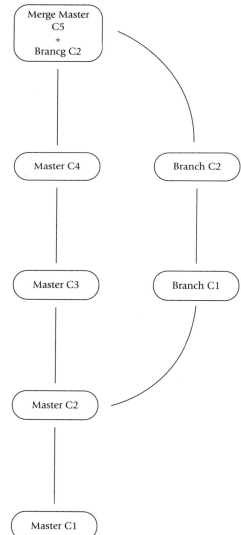

Figure 16.1. Example branch diagram.

Figure 16.2.
GitHub Desktop History tab. *GitHub Desktop* (2.2.4), GitHub (2019).

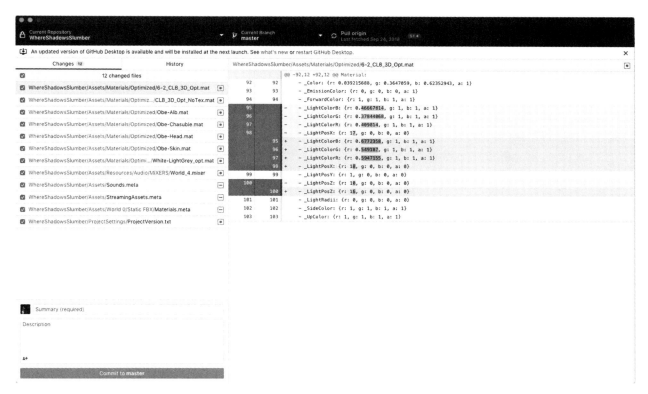

Figure 16.3.
GitHub Desktop Changes tab.

the "Changes" tab, which shows all of the local changes made to the working copy of the game in the local repo. On the right side, we can see the local changes that will be applied to the master. If we wanted to save these changes to the master, we would write a comment describing them in the comment box at the bottom left, then press the "Commit to master" button (below the comment box).

In this case, however, we want to "Pull origin," meaning we want to receive the changes from the master that others have made since we last worked on the game. However, when we click "Pull origin," we receive the following error (Figure 16.4), which tells us that our local changes would conflict with the new changes. If we were to pull at this time, the uncommitted local changes would be overwritten and lost.

To resolve this issue, we must therefore first deal with our local changes by either choosing to commit them and then to pull from the newly updated master, save them elsewhere for later, or simply delete them. In this case, we do not want to save any of these local changes, so the solution is quite simple. To discard them, we highlight them, right-click, and select "Discard all changes" (Figure 16.5). We are now finally able to successfully "Pull origin" and have our local repo conformed with the master.

We are finally up-to-date and ready to begin today's work on the game. In this case, we end up making some edits to the World 0 mixer, changing some of the volumes of different mixer groups. Now that we have made these local edits, we can see them listed in the "Changes" tab of the *GitHub Desktop* UI (Figure 16.6).

After commenting on these changes with a note clearly describing what they are, we click "Commit to master" (Figure 16.7).

We can now see in the top right of the UI window that the changes have been successfully committed and are now ready to be pushed to the master. To do so, we click "Push origin." These changes will be pushed to the server and will show up in the History section, which displays past actions (Figure 16.8).

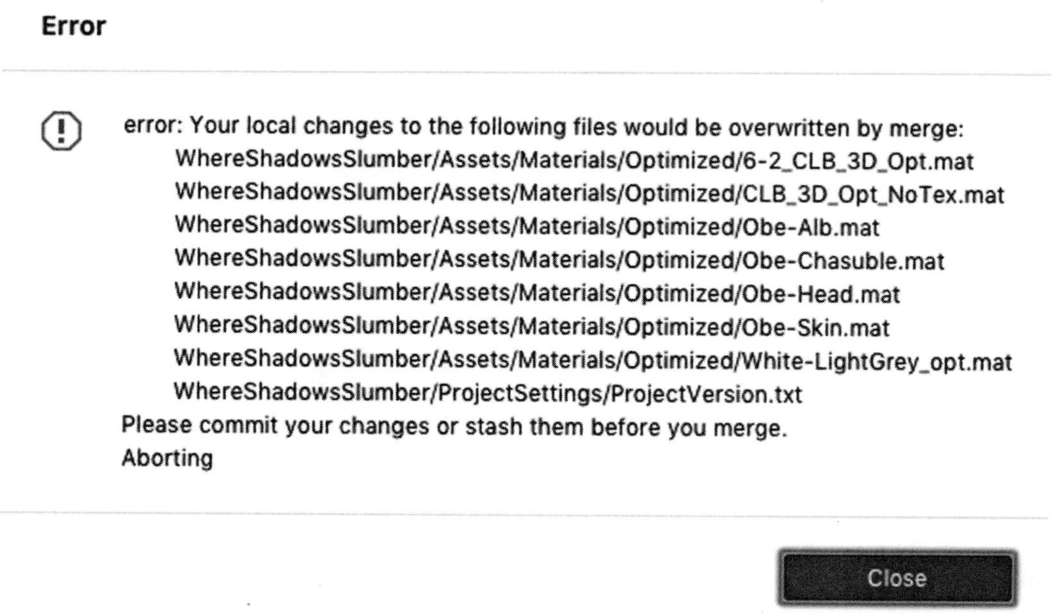

Figure 16.4.
GitHub Desktop example error message on "Pull origin."

Error

error: Your local changes to the following files would be overwritten by merge:
  WhereShadowsSlumber/Assets/Materials/Optimized/6-2_CLB_3D_Opt.mat
  WhereShadowsSlumber/Assets/Materials/Optimized/CLB_3D_Opt_NoTex.mat
  WhereShadowsSlumber/Assets/Materials/Optimized/Obe-Alb.mat
  WhereShadowsSlumber/Assets/Materials/Optimized/Obe-Chasuble.mat
  WhereShadowsSlumber/Assets/Materials/Optimized/Obe-Head.mat
  WhereShadowsSlumber/Assets/Materials/Optimized/Obe-Skin.mat
  WhereShadowsSlumber/Assets/Materials/Optimized/White-LightGrey_opt.mat
  WhereShadowsSlumber/ProjectSettings/ProjectVersion.txt
Please commit your changes or stash them before you merge.
Aborting

Working with Developers

Figure 16.5.
GitHub Desktop "Discard all changes" dialogue.

Figure 16.6.
Local edits reflected in the *GitHub Desktop* Changes tab.

Of course, since this was only a hypothetical exercise, we did not wish to actually push these changes to the server. In order to avoid actually pushing these changes, a reverse commit was used prior to the push. Essentially, this push changed only the record in the history, but not the game itself. It shows that changes were first made, then reversed, so nothing in the master actually changed other than the historical record of this work, as seen in Figure 16.9.

While this section on version control has offered you a brief introduction to how it functions and why it is used, there are numerous resources online that will provide you

Implementing the Score

Figure 16.7.
"Commit to master" in *GitHub Desktop*.

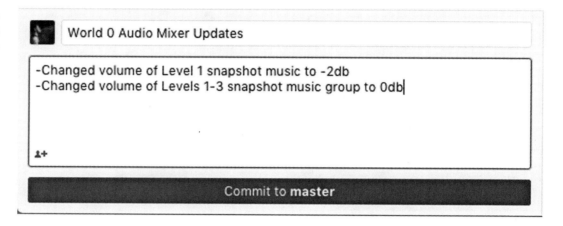

Figure 16.8.
A populated History tab of *GitHub Desktop*.

Figure 16.9.
The record has changed in the *GitHub Desktop* History tab.

with detailed knowledge of its functionality. You should choose a popular version control system and begin experimenting with it in order to gain some real-world practice and experience. It is highly likely that you will end up using version control when you work on games. In fact, doing so is highly recommended as it will save you and your team quite a bit of time!

Audio Optimization

While most of us are constantly using software, it is easy to forget that each program we use functions by utilizing the hardware built into our computers. Part of what makes a program stable and efficient, thus allowing it to provide an enjoyable user experience, is how well the programmers craft the interaction between this software and the computer's built-in components. Optimization is the process of compressing (essentially, strategically shrinking) files and determining how they load into the game so that the game functions as efficiently as possible. Since audio files tend to be one of the largest asset types in games, optimizing a game's audio is an extremely important part of the process. Audio can single-handedly make a game 10 times too large or inefficient if it is not handled properly. The optimization process all depends on how much power your computer or device has available. In order to properly carry out the procedures, you need to have an understanding of how a computer's CPU, RAM, and hard drive all work together.

Understanding Computer Memory and CPU

RAM stands for *random access memory*. A computer's RAM is essentially its short-term memory. Much like a human, it can store a certain amount of information to be called upon and utilized in the near future. This information is quick and easy to access, making it exceptionally useful for storing items that we know we will use a number of times. The amount it can store depends largely upon its size. The better your short-term memory, the more multitasking you can perform without becoming overburdened, and the same is true of your computer. More RAM, therefore, means more can be stored in short-term memory, making the access much faster and decreasing how often the computer needs to draw information from the hard drive.

You can think of the *hard drive* as the computer's long-term memory. It stores the majority of the information, but that information is often more time-consuming to access. When a specific item is needed, the computer grabs it from the hard drive and brings it into the CPU to be processed.

CPU stands for *central processing unit*. This computer component handles the processing of all the information. This information is taken either from the RAM or the hard drive. Information stored in the RAM, of course, is more readily available and thus more quickly processed by the CPU.

In summary, these three essential computer components work together to access and process information. The larger the files, the more time is required to access each informational element. If data is stored in the hard drive, or long-term memory, it takes the longest to access it. If it is in the RAM, it can more quickly be brought to the CPU to be processed, but using up too much memory also adds strain and affects a computer's ability to perform. For these reasons, we need to think about how much weight to put on each component by determining the types and sizes of files, as well as where to load them from. This process is called—you guessed it, optimization.

Optimization

Because all hardware has its limitations, all games must be *optimized* to function as best as they can for that hardware. The vast majority of files in the raw version of a game are far too big to use when the game is actually built, requiring an unreasonable amount of power to load. Therefore, these files, including textures, animations, and of course audio, are all compressed into smaller sizes and specified to load in a specific manner to best utilize the available hardware power so that the game runs as smoothly as possible. As an indie composer, it is important that you understand how this process works, because there is a high likelihood that it will be your responsibility to ensure that the audio is properly optimized. Often, you will be given an audio budget, a certain limitation to how much space the audio can take up and how much processing power it can use. It will be your job to ensure that the audio is using as little space and power as needed to run, and that it stays within these budgetary limits.

Optimization Settings in *Unity*

When you import a clip in *Unity*, you are given a variety of import settings in the inspector. The way you set these determines how these files are loaded at runtime, where they are stored, what type of compression is used to shrink them, and how much is applied. As you can see in the following images, the original sizes of the imported audio clips are significantly larger than the imported sizes. This is because the import settings are applied and the files are compressed to be better optimized for use in the game.

Load Type

The "Load Type" setting gives you the power to determine how each file is loaded into the game. This gives you an important level of control over which piece of computer hardware is used for each file.

When you set audio files to "Decompress on Load," they are stored compressed into smaller sizes, which allows them to take up less disk space (Figure 16.10). They are then decompressed the moment the sounds are loaded into the game or level, or rather, when the level is loaded. This setting is great for small, short audio clips like sound effects because these small files can be quickly and easily decompressed at the start of a level without causing a significant amount of lag. With this in mind, know that the decompression process uses more memory than keeping the file pre-compressed in memory. Therefore, decompressing a large, longer audio file on load would be an inefficient use of memory and can be more costly.

When a file is "compressed in memory" (Figure 16.11), it is stored in the RAM and is not decompressed until the instant it is needed within the gameplay. In many cases, larger files are too costly to decompress on load, so storing them in the memory until the time they are actually used can sometimes be a good solution. Often, this setting is a good choice for music and ambient loops that are a bit longer.

Setting a file to "streaming" (Figure 16.12) decompresses the sound from the drive directly. Therefore, it requires the least amount of memory, but can also be significantly

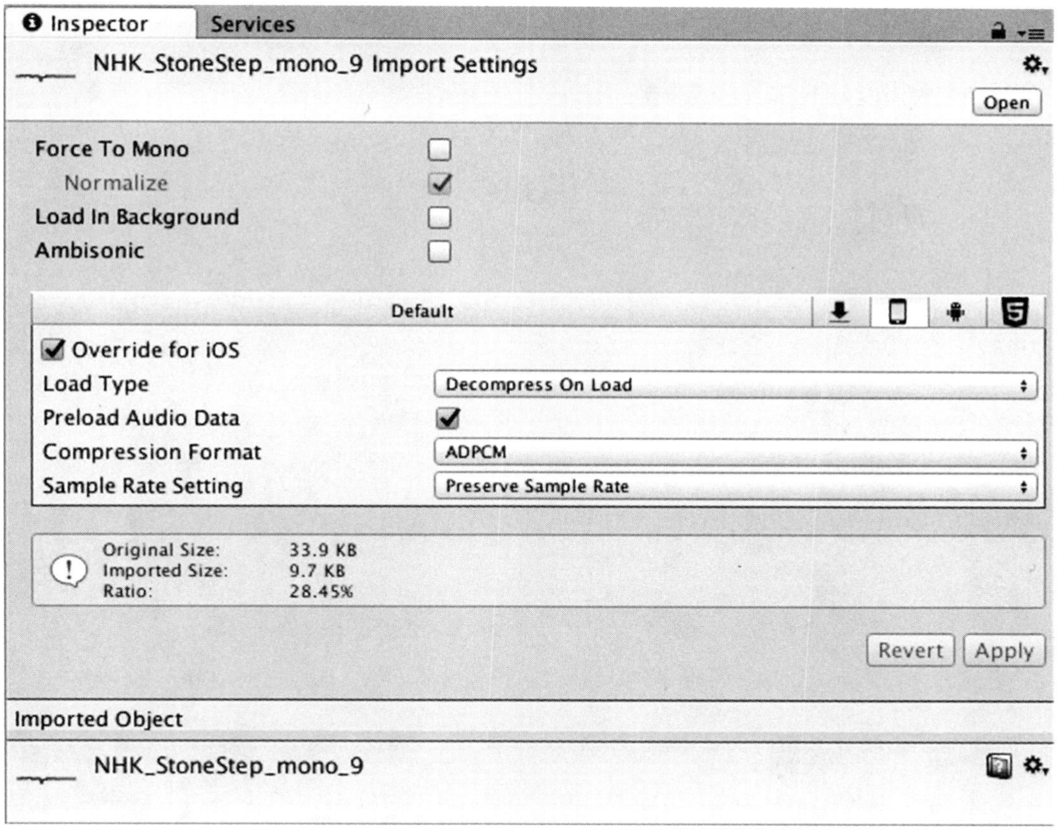

Figure 16.10.
Audio files set to "Decompress on Load" take up less disk space, as they are compressed until use. *Unity*, Unity Technologies (2019).

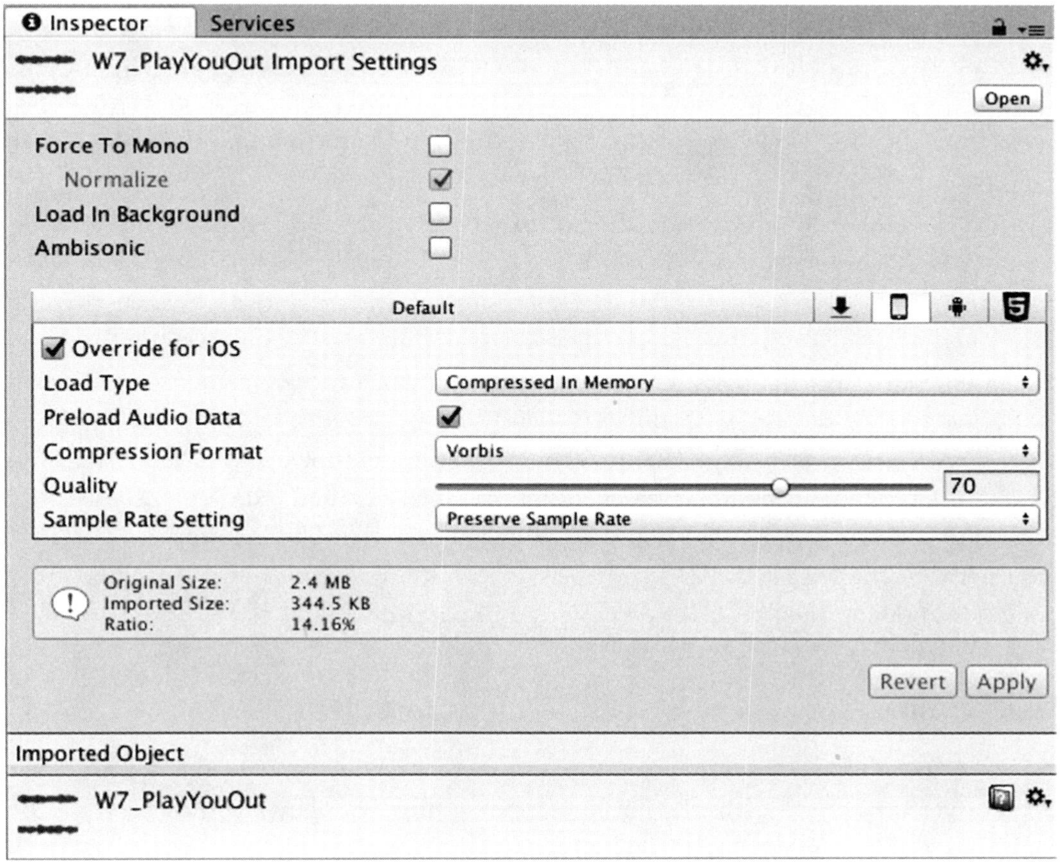

Figure 16.11.
Audio files set to "Compressed in Memory" is stored compressed in RAM.

Implementing the Score

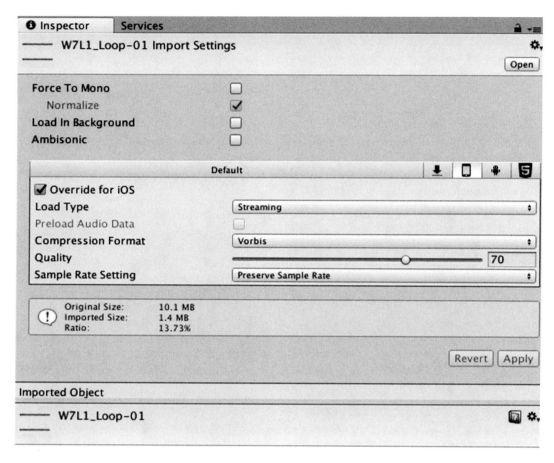

Figure 16.12. Audio files set to "Streaming" take up the least amount of RAM, but the most amount of CPU power.

more costly to the CPU. It is essential that you do not overuse streaming in any one given level as this can put unnecessary strain on your drive. However, this setting is an excellent solution for games that are using a great deal of RAM and do not have much to spare.

The amount that you can and should use of any of these three load types in combination with compression is, of course, dependent on how much memory and CPU budget are reasonable for the audio to use for any particular game. In general, particularly when working on mobile games, there are strict limitations to how costly the audio processing can be. In the case of *Where Shadows Slumber*, a mobile game released across numerous mobile devices with different hardware specifications, we aimed to use no more than 50 megabits (mb) of memory for the audio, and to keep CPU usage as low as possible. As mobile devices become more and more powerful, the limitations will likely decrease. However, developers often use the increase in available memory and CPU for the improvement of graphics and gameplay, so you must always know what your audio budget is, and you must be able to stay within those limits.

Target Quality

When streaming files or compressing them in memory, you will have the option to select a target quality for each file. This is a fader setting that allows you to decrease the quality to a certain percentage. Generally, you can decrease a file's imported quality to somewhere between 70 and 80 percent quality without any audible degradation; however, it will be up to you to use your ears to judge how small you are able to make a file before it begins to affect its sonic integrity.

Monitoring Your Optimization

Thankfully, most programs contain some sort of window that allows you to observe the amount of memory and CPU you are using, as well as details about the different types. In the case of *Unity*, this window is called the "Profiler." The *Where Shadows Slumber* level shown in the profiler here (Figure 16.13) is using a total memory of 47.6 megabytes (MB), staying just within our 50 mb threshold.

An important consideration is that the audio effects in the mixer use digital signal processing (DSP) to function. DSP takes up quite a bit of processing power as well. You'll notice in the profiler that the total audio CPU usage is 6.5 percent. In this case, we had a single reverb running, as well as some compression. However, this is a fairly

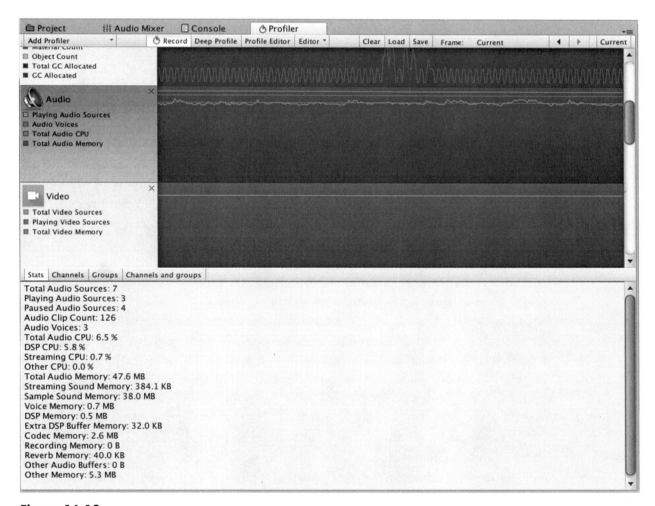

Figure 16.13.

Unity Profiler window example from *Where Shadows Slumber*.

small amount of audio effects. Be sure to keep an eye on how any real time digital signal processing is affecting your optimization.

In the end, our uncompressed audio folder in the *Unity* project was a whopping 1.4 gigs. Our goal was to keep the entire game under 500 MB in size. After applying all of our compression settings and building the game, compressing most of our streaming files down using the Vorbis compression format at around 70 percent quality, we checked the size of the audio. The 1.4 gigabytes was converted down into 200 MB without losing much audible quality, an acceptable size.

Being Smart with Resources

Aside from compression settings, another important part of memory optimization is ensuring that you are not using more audio elements than necessary at any given time. Looking at the profiler in Figure 16.13, you can see that the "Total Audio Sources" available in the scene is seven, while only three of them are actually playing. You will also notice that there is a total of 126 audio clips loaded into the scene. In *Where Shadows Slumber*, our audio objects, including the `MusicManager`, `SoundManager`, and `FXMachine`, are loaded per world, as opposed to per level. This means that all of the necessary audio clips are present for every level in that world.

Loading our audio objects per world was advantageous because it allowed us to create seamless audio transitions between levels because the audio itself does not need to be loaded each time a new level is entered. However, if we had been using too much memory and space, we might have optimized better by having a separate `SoundManager` loaded for every single level, allowing us to load significantly fewer audio clips and save processing power.

It is also important to turn off unused resources. Early in the process of building our `FXMachine`, we ran into problems because we had not yet built a way of pausing unnecessary ambient tracks. All of the ambiences of a given world played at the same time, with those unneeded in that level simply muted. In such a case, there were seven audio sources, and all seven of them were playing, even if only three of them were heard. This took up significantly more resources than necessary, so we quickly built a way of pausing irrelevant audio tracks for each level. Note that in the profiler screenshot, only three out of the seven available audio sources are actually playing.

Nonetheless, loading all 126 audio clips for every single level is also unnecessary. Thankfully, we didn't run into any issues on this front, which would have meant rethinking how our audio elements load in the game. However, ensuring that only the clips necessary for each level were loaded would have been yet another way of optimizing the game's audio. Using *Wwise*, for example, one can easily create separate sound banks (collections of audio clips) and load only those necessary at any given time.

In many games, you may need to utilize significantly more voices ("voices" simply refers to audio elements that are currently playing) all at once. For example, a machine gun sound effect might fire so rapidly that each gunshot uses a separate audio source to play, making it quite possible to quickly rack up an exorbitant amount of voices. In such cases, you should set limits to how many voices can be used by a single sound element. Most game engines and middlewares, including both *Unity* and *Wwise*, also offer prioritization systems, allowing you to limit the total voices available and to give certain audio priority in cases where the limit is reached. This ensures that the most important audio is never lost because there aren't enough audio sources available.

In the end, each system you work with will have different hardware capabilities, and each game will utilize the available resources in different ways. Be sure to always think ahead and monitor how many resources the audio is taking up. This will save you a headache when you reach the optimization process of development.

Conclusion

In this chapter, we have explored two extremely important technical issues that arise on almost every project you will work on. When working with a development team, version control is an essential skill to have. Being able to use it comfortably will give you the freedom to work on a game with speed without having to bother your team for assistance with downloading the most current version. It will also ensure that your coworkers quickly receive any changes and updates you make. Optimization is another critical part of the development process, and often developers will not know how to optimize audio properly. Being able to do so yourself, therefore, is by far the most effective method for ensuring that the audio meets the budgetary requirements of the game while still maintaining great sonic fidelity. Be mindful of the available resources from the beginning to ensure a smooth optimization process when it is time.

Exercises

Exercise 1: Version Control

1. Create a GitHub account at GitHub.com.
2. Download the GitHub desktop app and sign in.
3. In GitHub Desktop, click the File menu and select "New Repository."
4. Fill in its name and description. Choose its local destination on your computer, and select *Unity* as its "Git Ignore" file (this file simply ensures that your uploads and downloads will function properly). Choose a license to protect your work if you'd like. Finally, click "Create Repository."
5. Open *Unity* and create a new project. Make sure you save it *within* your local repo folder.
6. Make some changes to the *Unity* file by creating a new scene and adding a couple of game objects.
7. Go back to the GitHub Desktop app and commit your changes with a detailed note about what they are.
8. Finally, push these changes to the master repo.

Exercise 2: Optimization

1. Open up a past *Unity* project.
2. Drag a large number of audio files into the game until the audio takes up around 200 MB of space. These files can be random audio files from your computer, or you can choose specific clips for populating the game's soundscape.

3. Create several new audio sources and drag some of these audio clips into the audio sources so that a number of them are playing at runtime.
4. Click play with the profiler open and observe how much memory and processing power the audio is utilizing. Write these numbers down or take a screenshot of the profiler.
5. Now, go into each audio clip's import settings.
6. Set short audio clips to decompress on load with ADPCM compression format.
7. Set longer audio clips (like ambience and music) to either streaming or compressed in memory, using the Vorbis compression format at around 70 percent quality.
8. Click play again and observe the differences in the profiler. How much memory and processing power is the audio taking up now? What else can you do to optimize this game's audio performance?

Notes

1. *Time Machine* keeps track of a hard drive in its many different states as the user makes changes by adding, deleting, editing, or moving files, preserving the *delta*, or change, from backup to backup. *Time Machine* can therefore be used to revert to an older version of the file system, or even to revert a single file back to a previous version of itself.
2. An interface that allows users to navigate and perform tasks on their computer with text-based commands.
3. Source Tree (3.1.2(216)), Atlassian (2019).
4. Online.
5. The folder where the game file is stored, otherwise known as the "repo."
6. Jackson Kelly (designer of *Where Shadows Slumber*), interview with author, April 14, 2019.
7. *Where Shadows Slumber*, mobile, developed by Game Revenant (New York: Game Revenant, 2018).
8. GitHub Desktop (2.2.4), GitHub (2019).

Epilogue
Game Music Contracts and Navigating the Business

Introduction

Though writing game music can be an incredibly fun and enriching experience, a career in game music can be highly competitive and difficult to navigate. This chapter explores the business side of creating game music, focusing primarily on the indie game market. We will learn about the different factors that go into negotiating a game audio contract, using examples provided by an attorney. We will also explore some practical elements of interpersonal networking for business.

Game Music Contracts

Formal contracts are rarely easy for the layperson to parse. Thus, in this section, we will use basic English to introduce you to the various elements of a typical indie game music score contract and potential negotiation scenarios, which should provide you with the tools you need to obtain fair rates for your work. Keep in mind that the author is a composer, not a lawyer. Any legal advice you find in this book does not take the place of what your attorney may tell you, nor shall the author be held accountable for any actions you may take based on this advice. The advice you may find here is for educational purposes only.

It is not uncommon for game audio budgets to be tight. This axiom is especially true in indie games, with budgets that are already small to begin with. In most cases, developers simply do not have the necessary financial resources to pay their composers up front, or even on a payment schedule. The simple act of asking for a full rate can sometimes have the unfortunate effect of scaring developers off. Fortunately, there are a number of different ways to receive compensation that are much more feasible for teams with small budgets and have the potential to yield a better result for you, the composer, in the future. Navigating these different payment scenarios will play a crucial role in your financial success.

What Makes a Good Rate?

Before discussing the different forms of compensation, we should discuss how much money you should ask for in the first place. This is a fundamental question that I have heard asked numerous times at various game audio conferences after nearly every lecture dedicated to the business of game audio. The answer is surprisingly complex, as there is no single correct way to calculate the worth of your audio work. Therefore, let us begin by considering some of the common ways composers may be compensated:

- *Per-minute*: Many composers charge a set fee per minute of music. This rate ideally factors in all of the time and production costs of creating that minute of music. In setting this rate, it might also be prudent to discuss anticipated additional production costs, like that required to record an instrumentalist, with the developers; frame this scenario as an optional (but worthwhile) addition to the overall quality of the final music product, explaining that such an action will, however, incur extra expenses.
- *Hourly*: Particularly common for when developers are unsure of how much music they might end up needing, you might instead consider simply charging per hour of your work. This would, of course, involve keeping a detailed log of your hours, but it is a fair method of charging for the total time you put into a project. Again, in this scenario, it might also be prudent to discuss anticipated additional production costs, as an expense beyond the fee for your work.
- *Flat fee*: Sometimes, a developer might state simply that "we have x amount available in our budget for the music." In this scenario, it may be better to simply charge a flat fee. Be very careful to factor in all of the work you anticipate doing and adding some buffer, as there will inevitably be more work to be done than you expect. Also, be sure that the estimated time you will be spending on this work keeps the fee "worth your time"; for example, you may enjoy earning $800 for one morning of work, but $800 for a month's worth of work may not be worth it to you. The flat fee will be paid for a certain amount of "delivery materials" as agreed upon in your contract. In this scenario, it is generally more difficult, but still worth exploring, the inclusion of additional production costs, should you decide the music could benefit from it.

When deciding how you wish to charge for your time and skill, there are a number of factors to take into account. How much time do you have available in your days to contribute to this project? How soon is the final deadline? How much work do you estimate it will be? In general, I have found this rule of thumb to be a good method: calculate how much work you think the project will require and double it. After all, chances are that you will run into a number of unforeseen difficulties along the way that make the project more difficult or time-consuming than you had originally predicted.

You must also take into account potential benefits beyond the immediate financial aspect when taking on project. For example, you might believe a great deal in the game, or you may wish to establish a relationship with a specific developer. You may want to build up your résumé, contribute art into the world, or hone your technical skills. In short, the benefits of working on a game often extend far beyond the money you might make up front—particularly if you negotiate a good contract—and you should approach each project with a wide view of the different forms of compensation available.[1]

Knowing all this, let us discuss some hard numbers. In 2020, $800 per minute of music is considered to be a fairly low rate when working on large projects with established game companies. "A-list" composers commonly charge upward of $1,500 per minute of music. However, such rates are often untenable for indie game developers' budgets. In fact, if you wish to score an indie game with a limited budget, you will need to be able to propose an "affordable" fee in order to entice the developer. For this reason, it may help to consider indie game projects partially as investments or strategic risks—all it takes is one good investment to pay off and make everything else worth it. Therefore, it is highly recommended that you negotiate a contract that includes future returns and benefits for your work, as opposed to simply being paid up front, especially if you are taking a fee cut from your normal rate. This future potential payoff usually comes in the form of *rev-share*, a percentage payment based on the game's gross revenue (or occasionally net, upon which it is more accurately termed *profit share*).

In most cases, it will be impossible for an indie developer to pay you a "fair" fee up-front, making the inclusion of percentage points a potentially excellent option for both parties. As an example, when PHÖZ (composer Alba S. Torremocha and I) worked on *Where Shadows Slumber* (2018),[2] we created around 50 minutes of music. We also created hundreds of sound effects and ambient layers, not to mention alternate versions of many of these. If PHÖZ had charged a per-minute rate of, say, $800, it would have cost the developer, Game Revenant, around $40,000. At this cost, they would have either not been able to hire us; or, if they were dead-set on working with us specifically, they would have had to hire us only for a highly limited set of delivery materials. While we knew working on the game would entail a great deal of work on our end, we were able to come to an agreement for a year's worth of audio creation that suited all parties by using a contract that detailed a number of different payment schemes. This did include direct payment installments as we passed project milestones; however, since the direct payment was far less than we had originally asked for, we also negotiated for *rev-share*.

If you anticipate that a specific project will require a great deal of work and feel you must therefore ask for a high price, you should be able to explain to the developers exactly why you are asking for that price. Many times, they will not initially understand why or how music would cost so much to create, so it is important that you approach the negotiation humbly, educating them about the amount of detailed and high-skill work that creating even just a minute of music entails.

As many of you will be working with indie game developers, the following is a detailed list and explanation of payment structures (or, rather, ways of being compensated for your music and audio). You can and should use a combination of these payment methods to develop a fair and equitable structure, tailoring the details to each particular project you may be working on.

Payment Structures

- *Full buyout/work for hire*: When the developers wish to own 100 percent of the copyright—the musical work (that is, the writer's and publisher's share of the music composition and, if applicable, the lyrics) and the sound recording—they will pay you an agreed-upon fee for your work that will constitute the full extent of your payment. In such cases, you should be careful to negotiate a price you feel is fair for the full amount of work you will put in, especially if you imagine the game might earn a large income (which you are ceding your rights to) in the future. Buyouts are more common in high-budget games, since the developers of those games are more likely to have the available funds. Since the rights involved are complex and can change depending on region, it is highly advisable to consult a lawyer before agreeing to such a contract point.

- *Payment schedule*: In many cases, developers will wish to pay according to a delivery schedule. This way, they do not have to pay such a large fee up front, especially if they have yet to approve the work you will produce. A payment schedule provides both a steady flow of income for you and a convenient method of payment for the developers.
- *Rev-share/profit-share*: As we have previously discussed, *revenue share* provides the composer with a percentage of the income of each game unit sold (as *profit-share* first subtracts other expenses). So, for example, if the game you have written music for is being sold on the Apple App Store, your share will be calculated *after* Apple takes its share (currently 30% of the total sale price). A fairly standard rate for indie game music rev-share is 10 percent; however, this number may be more or less, depending on how much music is provided, how much work is required, and how much the composer is being paid otherwise. Top indie composers will sometimes work for 15 percent, while those either just starting out or working on projects that do not require a great deal of work will work for only 5 percent—or even less. You should be aware of how much your audio is likely to contribute to the experience and immersion of the game. In many cases, the audio can completely change the gaming experience, meaning the mark you will leave is quite profound; you deserve a larger share. In other cases, the audio might be fairly minimal, a less-important aspect of the gameplay; you may "deserve" somewhat less. Keep this in mind when asking for a rev-share. Also, note that this is the riskiest form of financial compensation; if the game promises you 50 percent rev-share but sells one copy, it is far less profitable than a game that gives you 5 percent rev-share but sells 100,000 copies. Finally, note that profit-share is sometimes colloquially called rev-share, but since the amount is taken out after expenses like distribution or recoupment, a profit-share deal is always a worse financial deal for you than a true rev-share, which calculates from the gross. Try to negotiate for the latter when possible.
- *Bonuses*: If or when a game sells a certain amount of copies, a "bonus" payment may be triggered. Bonuses can be a useful negotiating factor. Rather than the developer having to pay the composer a rev-share for each copy sold (which could potentially mean that the composer is getting paid while the developer is still paying back debt), bonuses allow the developer to set a minimum threshold at which first bonus payment would then be received. For example, you might get paid a fixed initial fee of $5,000; however, if the game's revenue hits $20,000, you may be paid an additional $1,500; if the revenue then reaches $60,000, you may receive another $5,000; and so on. Using this type of structure shows the developer that you believe in the game and that you are willing to invest your time with the confidence that the game will be a financial success.
- *Soundtrack sales*: The game's soundtrack—the sound recording of the music—is yet another element that requires attentive negotiation. In your own contracts, you should consider including a clause that grants you the right to create the game's soundtrack under your name and distribute it. It is very rare for a game soundtrack to be a form of notable financial income; however, it can be quite profound in its effect on your career if it becomes popular or viral. For example, Disasterpeace's *FEZ* (2012)[3] soundtrack achieved a level of popularity that had a far-reaching positive impact on his renown and, thus, his career.
- *Musical copyright ownership/performance royalties*: In many cases, the developer might have no objection to your collecting the full writer's share and publisher's share of the

royalties. However, this is yet another negotiating point, as other developers will keep the publishing while you keep the writer's share. If you have not yet done so, you should register with your Performing Rights Organization (PRO) as both a publisher and a writer so that you have the infrastructure in place with which to collect both of these royalty types for your music—which is important because you want to get paid, for example, if the sheet music is ever sold. Also, a fair contract should grant the composer retention of the public performance rights. This way, if the music is ever performed publicly, the composer will be entitled to those royalties. As the streaming of games becomes more and more common, and as laws change surrounding online video streaming sites like YouTube and Twitch, owning your writer's share could prove to be an important source of income. Note that it is illegal in some countries to cede control of writer's share, so if you are unsure of your rights in this area, it is highly advisable to consult a lawyer before agreeing to such a contract point.
- *Ancillary uses*: Ancillary uses of music are those that fall outside of the originally intended use—here, in the game itself. Some examples of this might be: marketing for the game, such as in trailers or other promotional videos; spin-off games; sequels to the game; documentaries about the game; and so on. All of these cases can serve as additional opportunities to get paid. It is crucial to include a clause about ancillary uses in your contracts to ensure that your music is not exploited beyond its original intent without your receiving equitable additional income.
- *Company equity*: Similarly to offering a percentage of the game's income, some developers may be willing to offer part-ownership in their studio. This practice is most common among indie developers who are just starting out, perhaps even releasing their first game. In such cases, you might be able to (tactfully) point out that you have no pre-established reasons to trust that the game will indeed generate a future income; that you would like to be more than just an audio provider, but to be a full member of the production team; that a percentage ownership in the company would give you great inspiration and incentive to work regardless of the potential lack of dependable income.

Negotiating Payment Structures

Realistically, most indie contracts will involve putting most or all of these elements together into a single proposition for the developers. Most indie developers will not have sufficient funds to offer you a buy-out. Instead, you should be able to approach the developers with the distinct combination of the aforementioned payment methods you feel would be fair for the anticipated workload. In order to clarify how such a proposition might be put together, let us now take an example indie game and create a payment structure, outlining it in an equitable contract.

Example: Putting Together an Indie Game Contract

Thorn's Castle is a hypothetical 2D side-scroller in which the player must avoid being hit by arrows, cannonballs, and spears while jumping through levels of medieval towns and castles. *Thorn's Castle* has five different worlds, each with five levels. The developers have indicated to you that they would like a unique one-minute background music loop for each level. One minute of music per level and five levels per world means a total of 25 minutes of music.

Your first step is to analyze the gameplay and determine how much work will go into creating the game's soundtrack. In this case, the developers have already provided a rough vision for how much length of music they would like included in the game. However, they have not accounted for things like menu music, start screen theme music, or the complexity of musical arrangements necessary to complete the score.

In order to determine how much time creating the music will actually require of you, you must consider the sonic aesthetics of the soundtrack itself and the implications on the workload. The game's rustic, medieval landscape easily lends itself to an orchestral sound, which is what the developers have indicated they would prefer. From a compositional perspective, this aesthetic likely requires more work in order to produce the finished product than, say, a score of simple synth loops (though of course, this depends on your own skill set). You therefore anticipate that writing, orchestrating, and properly sequencing each minute of the score will take you around one day's worth of work. When thinking about your price per minute, you should calculate as if you would be paid one upfront fee with no back-end percentage, and also that the developers may very well request a number of revisions or even complete rewrites. A fair price for you, therefore, might be around $1,000 per minute of music. Thus, for the 25 one-minute music loops, as well as for the theme and menu music pieces, you determine that you should ask for somewhere between $27,000 and $30,000.

However, going into this project, you also know that the developers are on a rather limited budget. This is their first game, and the project is completely self-funded. The next question you must ask yourself, then, is *how important is working on this game to you?* Do you believe in its financial success, and/or are there benefits beyond the financial compensation? Do you like the developers as people and foresee the relationship being easy-going as opposed to high-stress?

After seeing the gameplay and spending some time with the developers, you determine that *Thorn's Castle* looks incredibly fun and engaging, and that you will likely very much enjoy working with them as people. Given how addictive you find the gameplay, you would be surprised if it did *not* experience some popularity (and, thus, financial success) in the indie market. You also feel it will be a gratifying artistic challenge to work on. Considering that you yourself are also in the early stages of your career, a game of high quality like this will provide an excellent portfolio piece for you, even if you do not make a lot of money.

Next, consider a few more important points. *Will you need to record any instruments?* If so, factor this cost (which should be a separate line item from your creative fee) into the contract after discussing it with the developers. *How quickly do they need you to produce the final product?* Will you be able to work on other projects as well, or will this be your sole livelihood for the months you spend working on it? In this scenario, the developers will need the music within two months, meaning you will be working on only their project for the duration of that time. In other words, for two months, your incoming cash flow would come solely from this game. You determine that despite the low budget, you would like to work on this game, to invest in (what you hope will be) the promising futures for both the game and your career.

With your knowledge of the developers' financial situation, you know that asking for $30,000 for the music is unrealistic. In fact, such a number might even scare them off completely, a result you are determined to avoid. After a great deal of thought, you determine that you could be willing to work on the game for only $15,000 of guaranteed income. A tactful approach for speaking to the developers about this number might be to first explain how you determined your original rate of $30,000 (pointing out the additional elements of menu music and main theme music), then indicating that, since you understand their budgetary constraints, you would be willing to work for only $15,000 up front. However, in order for this to be more equitable, you would like to also be invested in the game's future potential. Therefore, you also ask for a 10 percent net rev-share. You tell them that you are also open to negotiating other factors, and that you would be potentially very interested in a small equity ownership of their company and being a permanent member of their team moving forward into the future. You suggest that this might even be something they can decide after working with you for a time.

The developers get back to you. Politely, they say that $15,000 is still simply too much for them to afford at the moment, but they propose $6,000 to be paid to you on a schedule of $2,000 each time you complete nine tracks. They say that 10 percent also seems high for rev-share, considering that you would be the only team member being paid up-front before the game's release and that they have spent a great deal of money on the game already. However, they are open to offering you a percentage ownership of the company itself after the timely delivery of the music and if the relationship progresses positively. They propose a 5 percent revenue share only after the game has made back its expenses, as well as a potential 5 percent company equity upon your timely delivery of the "delivery materials" in the contract.

You have a lot to think about here. In this case, you are, in truth, guaranteed only $6,000 for quite a bit of work, a number substantially lower than your original "fair" projection. However, their openness to giving you a 5 percent company equity is generous, and the future implications are certainly exciting, as it means you will be a permanent member of their team. You are tempted to simply accept this offer based on that alone; however, with the short deadline in mind, you also have to think about short-term cash flow as you work on the game—after all, you also have to pay your rent and living expenses. You decide to offer a counter proposal you feel is reasonable. You ask for $8,000 according to the same payment schedule, an 8 percent revenue share only after the game's expenses are recouped, and the 5 percent company equity. They developers get back to you and agree to your deal!

Now that you have agreed on terms, you must ensure your contract has a clause that allows you to:

- be informed of the game's income and expenses so you can see when you will be paid and how much;
- review your equity ownership contract.

This exercise was, of course, a very specific example of a contract negotiation for one hypothetical low-budget indie game. In many cases, developers will be completely unwilling or simply unable to offer a part-ownership in their company. Their budgets and timelines will also all be different, so it will be up to you to determine what price you are willing to work for and to ensure that you negotiate an income you feel is equitable.

Next, let us discuss another important business skill: acquiring gigs in the first place!

Navigating the Business

Ultimately, the game industry is just like any other industry: everything revolves around relationships. Have you ever had the experience of seeing a colleague land an incredible gig and wondering why you did not get it, even though you feel you are equally or even more qualified? The fact of the matter is that while your skills are of great value *after* you land the gig—providing you with a platform for doing wonderful work and strengthening your relationships and connections—absolute skill level often plays little to no part in procuring a gig.

This is why the skill of *prospecting* becomes extremely important. Prospecting is the act of spending time looking for gigs, or developing "prospects." It is no simple task, and requires a great deal of time, patience, dedication, and organization. You must spend time each and every day opening up dialogues, following up with connections, showcasing and marketing your work, and strengthening relationships. One excellent method of prospecting is to simply spend the first one to two hours of every single day following up on leads, contacting new people and sending them your work, and having meaningful dialogues with them. Another important idea to keep in mind is that when you are experiencing a successful moment of your career and have a number of gigs lined up, you should not slow down prospecting. Only through this type of dedication will you maintain a steady freelance flow of work. So, where do you find developers to communicate with?

Where to Network

Face-to-face Networking Opportunities

Face-to-face networking, including online video chat meetings, is one highly effective way to create meaningful relationships with developers. The following is a non-exhaustive list of different places you can meet developers in person.

- *Regional organizations*: Without a doubt, you should search for and join game organizations in your area, as they often put together large events (like play tests or educational meetups). In New York City, the largest organization of game developers is *Playcrafting*. In addition to staging *PLAY NYC*, New York's largest video game convention, they host numerous other events throughout the year featuring local indie game developers and their work. These events, and events like these, are wonderful places to meet developers in need of game audio.
- *GANG*: GANG is short for the Game Audio Network Guild. With chapters across the country, the guild sets up regional events, which are excellent places to not only

learn about game audio, but also to meet other professionals in the game audio world. Indeed, one of the best ways to get game audio gigs is to make friends of your game audio colleagues, who will almost certainly pass gigs along to you when they are too busy to take them on themselves.

- *Game jams*: A game jam is an event that revolves around meeting a random group of game makers and creating a game over the course of only one or two days. Game jams are a great place to form meaningful professional relationships with those you work with. Jams usually begin with a prompt used to inspire the creation of a game. Then, people participating in the jam meet each other and form into teams. Often, there might even be a shortage of audio specialists, so you should make your skills known to the other developers around you. One game jam series of note, Global Game Jam, is a global event that takes place at different locations around the world. You should see if or when it is happening near you!
- *Educational institutions with game design programs*: Many colleges and universities offer game development majors at both the undergraduate and graduate levels. These students are often very amenable to the idea of having outside help for the audio. At New York University, for example, the Tisch Game Center hosts "Play-Test Thursdays"; every Thursday, all of the students—including many graduate students who will later go on to have fully funded projects—demo their in-progress games to the public. Once a year, the Tisch Game Center also hosts an event that showcases all of the games made by students that year. These types of events are an excellent way to meet early-career developers who are working on passion projects. While many of these students will not have much of a budget to speak of, if you are new to the game music industry, it can be a fantastic way to get started working on games.
- *Classes*: If you are new to the game music industry or eager to branch into game programming, you might even consider taking a game programming or design class, either online or at a local school. In addition to garnering knowledge and developing valuable skills yourself, taking such a class is an incredible way to establish friendships and work on projects with aspiring developers in a setting that requires you to interact with them on a regular basis.
- *Meetup.com*: Meetup.com is a website that facilitates the organization of local meet-up events. There are numerous game developer events on the website. All you have to do to participate is to create a free account and search for game-related meetups near you. You will certainly find some events with their dates and times listed. RSVP, show up, and mingle with a number of other professionals in your field.
- *Conferences*: There are numerous conferences throughout the world that center around the showcasing of games. Some of these are better than others, however, when it comes to networking. The most prominent conference in the United States is the Game Developers Conference, or GDC, which is held annually in March in San Francisco. While attending this conference can indeed be expensive, it is one of the single-best networking opportunities available because it is geared toward professionals instead of fans. However, it is also easy to get lost in its "grandeur." There are also many other smaller conferences held throughout the world; you should research which are being held near you.

Additional Online Networking Opportunities

The following is a list of online locations you can use to find developers and later set up face-to-face conversations.

- *Mobile stores*: Go to the Apple App Store or the Google Store and search for games with the word "demo" or "beta" in their titles. You should find a long list of games that are still in their demo stage. You can download these demos and take note of whether they have audio. If they do, consider whether it sounds professional, and do your research to find out whether the developers have already hired audio creators. If not, this is an excellent opportunity to (respectfully) get in touch with them, offering your feedback and your vision of the game's potential while letting them know that you are interested in working with them.
- *Crowdfunding platforms*: Many indie games use crowdfunding platforms like Kickstarter or Indiegogo to raise money. Since audio is often added in the latter stages of development, many of these games do not yet have professional audio integration and may even be raising part of their funds in order to hire a composer and/or sound designer. This is an excellent opportunity to get in touch with developers to find out if your services could be useful to them, particularly since you know they will have some budget (and even what that budget might be).
- *Game demo websites*: There are a number of websites with dedicated followings that are used to upload and distribute game demos. Developers use these websites to find beta testers for their games and receive valuable feedback on those betas. You can sign up to be a beta tester on these websites, contributing valuable testing information to games that are in the middle of development, as well as finding games that may be in need of professional music.
- *Meetup.com*: This website can also be useful for purely online prospecting efforts. Even if you cannot attend an event, you will still be able to access the guest list. You can scour this list, doing research about those who are attending the event, discovering whether they are developers and, if so, whether they are currently working on any projects. Using this information, you could get in contact with them, offering your services with a thoughtful explanation of how you came across their work and why you feel you would be a good fit for their project.
- *Social media*: Any of the mainstream social media platforms are littered with game developers who are beginning the marketing for their games well in advance of release. You will, without a doubt, be able to find numerous opportunities to get in touch with budding developers who are in need of your skills. You can begin by searching for relevant hashtags, for example, "#indiedev," "#indiegame," or "#mobilegame," to name just a few.

Ultimately, the best way to pick up projects is to be diligent about your networking efforts. The more events you can attend, whether in-person or online, the better results you will achieve. The more you come across as a bringer of solutions instead of a bringer of problems, the better results you will achieve. To this end, you should avoid coming across as someone solely interested in self-promotion. You not only are more likely to make meaningful connections, but also will have a better networking experience if you attend these events with the primary purpose of making friends. Walk through the door, introduce yourself to people, and if you feel a good connection, spend ample time developing a meaningful relationship with them. Even if they are not working on any exciting projects currently, you never know what they may be working on in the future, or to whom they will introduce you down the road. Be humble, spend time prospecting every day, and remember that anyone you meet has the potential to open a career-changing door for you.

Exercises

Contracts
1. Use an online resource to find a real indie game you would like to work on.
2. Using the video content available on their social media, consider a music design for their game. Determine how much work you anticipate this requiring.
3. Think through the different payment structures you feel would be fair for creating audio for the game. Put together a basic payment structure you would use were you to communicate with the game developers.

Networking
1. Find out if there are any game jams or conventions you can be a part of. If there are, pencil them into your calendar and sign up to attend!
2. Using online sources, track down three different games in their early stages and get in touch with the developers, giving them valuable feedback and thoughtful explanation of how you think you could contribute to their projects. Remember to keep your message brief and to the point, but with meaningful content.

Supplementary Materials

The following contract has kindly been drafted by Joe Frateschi, an attorney licensed to practice in New York State, for your use. You may use it as is, or use it as a template for creating your own contracts for each project. You may download both .PDF and .docx versions from the companion website so that you can modify this contract to suit the needs of each project. ▶(17.1)

AGREEMENT

THIS AGREEMENT, made and entered into this ___ day of _____, 20__, is by and between [_____] (the "Developer"), with a mailing address of [_____], and [_____] (the "Composer"), with a mailing address of [_____].

RECITALS:

WHEREAS, Developer desires to license a musical composition from Composer for use with a game product that Developer intends to market; and

WHEREAS, Composer is willing and able to license such musical composition in accordance with the terms recited herein.

NOW, THEREFORE, for good and valuable consideration, the parties hereto agree as follows:

1. SCOPE OF WORK Composer shall create and provide to Developer the work described on Schedule A (the "Work") in accordance with the criteria set forth therein, which is incorporated herein and made a part hereof, for inclusion in that certain game entitled [_____] (the "Game").

2. GRANT Subject to the terms hereof, Composer hereby grants to Producer, its successors and assigns, the exclusive license to do any or all of the following in the entire world (consisting of planet Earth) (the "Territory"): (a) To record, reproduce and fix the Work in synchronism or timed relation with the Game; and (b) To make, reproduce, publicly perform and display, or distribute copies of the Work, as contained and synchronized with the Game, by any means or medium now known or hereafter to become known.

3. RESTRICTIONS Excepting the rights explicitly granted to Developer in Section 2 above, this Agreement does not grant the Developer the right to: (a) make any change to the Work not expressly authorized in writing by Composer; and (b) make any other use of the Work other than as expressly authorized herein.

4. TERM Developer shall have the rights granted hereunder in perpetuity.

5. OWNERSHIP Composer is and will be the sole and exclusive owner of all right, title, and interest in and to the Work, including all patents, copyrights, trade secrets and other intellectual property rights therein.

6. RESPONSIBILITIES Composer agrees to create, develop, and provide Work to Developer in accordance with the delivery schedule provided for in Schedule A attached hereto and made a part hereof (the "Delivery Schedule"). Developer shall accept the Work as delivered by Composer in accordance with the Delivery Schedule subject to the following rights: (a) Developer may request full rewrites (i.e., substantive changes) to [_] of the items contained in the Delivery Schedule; and (b) Developer may request revisions (i.e., non-substantive changes) of [_] of the items contained in the Delivery Schedule.

7. COMPENSATION
    A. In consideration of the license granted by Composer hereunder, Developer shall pay to Composer the following flat fees: (a) [$_____] upon the execution of this Agreement by the parties; (b) [$_____] upon the timely delivery to and acceptance by Developer of one half (1/2) of the items contained in the Delivery Schedule; (c) [$_____] upon the timely delivery to and acceptance by Developer of all of the items contained in the Delivery Schedule.
    B. As additional consideration for the license granted by Composer hereunder, Developer shall pay to Composer, the following royalties: [Text percentage rate of net sales] percent ([percentage rate of net sales]%) of the net sales income received by Developer or its licensees from the sale by Developer

of the Game. For purposes of this Agreement, the term "net sales income" will mean gross monies actually received by Developer for the Game, less the following costs directly paid and actually, demonstrably incurred (or debited against Developer): (a) costs and expenses in connection with the manufacturing, sales, advertising, distribution and shipping of the Game.

8. ACCOUNTING OF ROYALTY
    A. Developer shall account to the Composer within ninety (90) days of the end of each accounting year and will provide the Composer with a statement of accrued remuneration earned during the preceding accounting period and will keep proper books of account and supporting statements in respect of the Work, and will pay to the Composer all royalties due to the Composer (if any). All statements and payments shall be sent to Composer's address above.
    B. All royalty statements and other accounts rendered by Developer to the Composer shall be binding upon the Composer and not subject to any objection by the Composer for any reason unless specific objection in writing stating the basis thereof is given within two (2) years from the date rendered. The Composer will have the right to inspect and make extracts of the Developer's books of account in relation to the Work through the Composer's duly appointed advisers at an office designated by the Composer during regular business hours, no more than once a year on giving thirty (30) days' notice to Developer.

9. COMPOSER'S REPRESENTATIONS AND WARRANTIES

Composer hereby represents and warrants that the Work:
    A. is Composer's sole and original creation or that and permissions have been obtained from any third parties as necessary for Developer to use the Work as contemplated by Developer;
    B. has not been, and prior to Developer's use or publication thereof will not be, published by or sold to another party, in whole or in part, without written consent from Developer; and
    C. does not, and use thereof as contemplated by Developer will not, infringe or otherwise violate any right of any third party, including any copyright, trademark, patent, trade secret, or other intellectual property right, or any right of publicity or privacy.

10. SCREEN CREDIT Composer shall be credited in the Game as: Music by [_____].

11. MISCELLANEOUS
    A. This Agreement is personal to Composer. Composer shall not assign or otherwise transfer any of its rights, or delegate, subcontract, or otherwise transfer any of its obligations or performance, under this Agreement. Developer may freely assign or otherwise transfer all or any of its rights, or delegate or otherwise transfer all or any of its obligations or performance, under this Agreement. This Agreement is binding upon and inures to the benefit of the parties hereto and their respective permitted successors and assigns.

B. This Agreement shall be governed by and construed in accordance with the internal laws of the State of [insert state].

C. This Agreement constitutes the entire agreement of the parties with respect to the subject matter contained herein, and supersedes all prior and contemporaneous understandings and agreements, whether written or oral, with respect to such subject matter. This Agreement may not be changed or modified, except by an instrument executed by the parties.

IN WITNESS WHEREOF, this Agreement has been executed by the parties as of the day and year shown below.

DEVELOPER:

[Insert Corporate Name]

Dated: _____, 20__

By:

_____

Name:

Title:

COMPOSER:

Dated: _____, 20__

_____

Notes

1. One good rule of thumb, though, is to never work for "exposure." Often (though not always), projects with budgets for music, from established or exciting developers, are the projects that also get the exposure with measurable worth.
2. *Where Shadows Slumber*, mobile, developed by Game Revenant (New York: Game Revenant, 2018).
3. *FEZ*, game disc, developed by Polytron Corporation (Montreal: Trapdoor, 2012).

BIBLIOGRAPHY

Abramson, Zach, and Ronny Mraz (*Just Cause 4* music team). Interview with author, April 1, 2019.

Adventure Time: Magic Man's Head Games, download, developed by Turbo Button. (=Atlanta, GA: Cartoon Network, 2016.

"Al Alcorn Interview." Cam Shea, IGN, published March 10, 2008, updated May 12, 2012, accessed December 16, 2019, https://www.ign.com/articles/2008/03/11/al-alcorn-interview?page=1.

Asteroids. Arcade, developed by Atari. Sunnyvale, CA: Atari, 1979.

Barnes, Craig (composer of *Sound Sky*). Interview with author, February 26, 2019.

Bendy and the Ink Machine: Chapter Two: The Old Song. Download, developed by theMeatly Games. Ottawa, ON: theMeatly Games, 2017.

Benis, Mark (composer of *The Worst Grim Reaper*). Interview with author, April 25, 2019.

Bioshock Infinite. Game disc, developed by Irrational Games. 2K Games, 2013.

Black Ops III. Game disc, developed by Treyarch et al. Santa Monica, CA: Activision, 2015.

Braid. Game disc, developed by Number None. Redmond, WA: Microsoft Game Studios, 2008.

Buckner and Garcia. "Pac Man Fever." Single, Columbia/CBS Records, December 1981, 7-inch.

Byrd, John. "Introduction to Machine Learning for Game Audio." GameSoundCon 2018 (convention lecture), Millennium Biltmore Hotel, Los Angeles, CA, October 10, 2018.

Call of Duty: WWII. Game disc, developed by Sledgehammer Games. Santa Monica, CA: Activision, 2017.

Celeste. Download, developed by Matt Makes Games. Vancouver, BC: Matt Makes Games, 2018.

Chang, Larry Yucheng (VR audio specialist). Interview with author, April 17, 2019.

Cheng, William. *Sound Play: Video Games and the Musical Imagination*. Oxford: Oxford University Press, 2014.

Collins, Karen. *Game Sound: An Introduction to the History and Practice of Video Game Music and Sound Design*. Cambridge, MA: MIT Press, 2008.

Covenant, James. "The Music of Super Mario 64 | Game Music Archaeology Ep. 1." YouTube video, 16:21, March 2, 2017, http://www.youtube.com/watch?v=PE7r5trvOUI.

Dark Souls III. Game disc, developed by FromSoftware. Tokyo: Bandai Namco Entertainment, 2016.

Dead Rising 4. Game disc, developed by Capcom Vancouver. Redmond, WA: Microsoft Studios, 2016.

Desiderio, Erik. "Composing Music for VR Games: *Adventure Time* Case Study." Game Developers Conference 2016 (convention lecture), Moscone Convention Center, San Francisco, CA, March 16, 2016.

Díaz Gasca, Juan S. *Music beyond Gameplay: Motivators in the Consumption of Videogame Soundtracks*. Mount Gravatt, Australia: Griffith University Press, 2013.

Dino Polo Club. "Mini Metro." Published 2016, accessed October 9, 2019, https://dinopoloclub.com/games/mini-metro/.

Dolby, Thomas. "The Next Generation of Non-linear VR Composers." GameSoundCon 2018 (convention lecture), Millennium Biltmore Hotel, Los Angeles, CA, October 10, 2018.

Bibliography

Donkey Kong Country 2: Diddy's Kong Quest. Game cartridge, developed by Rare. Kyoto: Nintendo, 1995.

Dungeon Siege. Game disc, developed by Gas Powered Games. Redmond, WA: Microsoft, 2002.

Edwards, Michael. "Algorithmic Composition: Computational Thinking in Music." *Communications of the ACM*, July 2011, cacm.acm.org/magazines/2011/7/109891-algorithmic-composition/fulltext.

Electroplankton. Game cartridge, developed by indieszero. Kyoto: Nintendo, 2005.

Eng, Kristofer, and Philip Bennefall. *Elias*. Stockholm: Elias Software, 2019.

Ermi, Laura, and Frans Mäyrä. "Fundamental Components of the Gameplay Experience: Analyzing Immersion." *Changing Views: Worlds in Play—Selected Papers of the 2005 Digital Games Research Association's Second International Conference*, pp. 15–27. British Columbia, Canada: Vancouver, 2005.

Farpoint. Game disc, developed by Impulse Gear. San Mateo, CA: Sony Interactive Entertainment, 2017.

FEZ. Game disc, developed by Polytron. Montreal, Quebec: Trapdoor, 2012.

Final Fantasy. Game cartridge, developed by Square. Tokyo: Square, 1987.

Final Fantasy VI (Final Fantasy III in North America). Game cartridge, developed by Square. Tokyo: Square, 1994.

Final Fantasy VII. Game disc (2), developed by Square. Tokyo: Square, 1997.

Flower. Game disc, developed by Thatgamecompany. Tokyo: Sony Computer Entertainment, 2009.

Giants: Citizen Kabuto. Game disc, developed by Planet Moon Studios. Los Angeles, CA: Interplay Entertainment, 2000.

GitHub Desktop (2.2.4). GitHub, 2019.

Gould, Richard. "The Programmed Music of 'Mini Metro': Interview with Rich Vreeland (Disasterpeace)." *Designing Sound*, published February 18, 2016, accessed May 7, 2019, designingsound.org/2016/02/the-programmed-music-of-mini-metro-interview-with-rich-vreeland-disasterpeace/.

Grace, Krista. "Music and VR: Interview with Farpoint PlayStation VR Composer Stephen Co." *Edgy Universe* (blog), May 27, 2017, edgy.app/farpoint-PlayStation-vr-stephen-cox-interview.

Grand Theft Auto: San Andreas. Game disc, developed by Rockstar North. New York: Rockstar Games, 2004.

Guild Wars 2. Download/online, developed by ArenaNet. Pangyo, South Korea: NCSoft, 2012.

Guitar Hero. Game disc, developed by Harmonix. Mountain View, CA: RedOctane, 2005.

Hanning, Barbara Russano, et al. *Concise History of Western Music*. New York: W. W. Norton, 2009.

Horizon Zero Dawn. Game disc, developed by Guerilla Games. San Mateo, CA: Sony Interactive Entertainment, 2017.

Hyper Light Drifter. Download, developed by Heart Machine. Culver City, CA: Heart Machine, 2016.

Inside. Game disc, developed by Playdead. Copenhagen: Playdead ApS, 2016.

"Intelligent Music Systems." Accessed November 4, 2019, http://www.intelligentmusicsystems.com.

Just Cause 3. Game disc, developed by Avalanche Studios. Tokyo: Square Enix, 2015.

Just Cause 4. Game disc, developed by Avalanche Studios. Tokyo: Square Enix, 2018.

Kingdom Come: Deliverance. Game disc, developed by Warhorse Studios. Hōfen, Austria: Deep Silver, 2018.

Kondo, Koji, and Ryo Nagamatsu. "Cavern Theme (Going Underground)." On *The Legend of Zelda: A Link between Worlds Original Soundtrack*, Nintendo 3679366, 2015, compact disc, 2013.

Kontakt 6. Native Instruments: Germany, 2019.

Lamperski, Phil (lead audio designer of *Rise of the Tomb Raider*). Interview with author, December 6, 2018.

Lehman, Frank. *Hollywood Harmony: Musical Wonder and the Sound of Cinema*. Oxford: Oxford University Press, 2018.

Levitin, Daniel. *This Is Your Brain on Music: The Science of a Human Obsession*. New York: Penguin, 2006.

Limbo. Game disc, developed by Playdead. Copenhagen: Playdead ApS, 2010.

Lundgren, Robert. Interview with author, May 16, 2019.

Mario Kart 64. Game cartridge, developed by Nintendo EAD. Kyoto: Nintendo, 1996.

Mario Party 2. Game cartridge, developed by Hudson Soft. Kyoto: Nintendo, 1999.

Mass Effect: Andromeda. Game disc, developed by BioWare. Redwood City, CA: Electronic Arts, 2017.

Masuda, Junichi. "Battle! (Wild Pokémon)," 1996. On *Pokémon Red & Pokémon Blue: Super Music Collection*, The Pokémon Company/OVERLAP OVCP-0006, compact disc, 2016.

Masuda, Junichi. "Final Battle! (Rival)," 1996, on *Pokémon Red & Pokémon Blue: Super Music Collection*, The Pokémon Company/OVERLAP OVCP-0006, compact disc, 2016.

Medina-Grey, Elizabeth. "Modular Structure and Function in Early 21st-Century Video Game Music." PhD dissertation, Yale University, 2014.

Merriam-Webster.com Dictionary. https://www.merriam-webster.com/dictionary/immerse.

Metal Gear Solid. Game disc, developed by Konami Computer Entertainment Japan. Tokyo: Konami, 1998.

Metal Gear Solid 2: Sons of Liberty. Game disc, developed by Konami Computer Entertainment Japan. Tokyo: Konami, 2001.

Michael Jackson's Moonwalker. Floppy disk, developed by Emerald Software et al. Witton, UK: U.S. Gold, 1989.

Minecraft. Download, developed by Mojang. Stockholm: Mojang AB, 2011.

Mini Metro. Download, developed by Dinosaur Polo Club. Wellington, New Zealand: Dinosaur Polo Club, 2015.

Mitchell, Briar Lee. *Game Design Essentials*. Hoboken, NJ: Sybex, 2012.

Moore, Michael. *Basics of Game Design*. Natick, MA: A K Peters/CRC Press, 2011.

Moss. Game disc, developed by Polyarc. Seattle, WA: Polyarc, 2018.

Mother 2 (*Earthbound* in North America). Game cartridge, developed by Ape, Inc., and Hal Laboratory. Kyoto: Nintendo, 1994.

Mutant Year Zero: Road to Eden. Game disc, developed by The Bearded Ladies Consulting. Oslo: Funcom Oslo AS, 2018.

"Nishikado-san Speaks." *Retro Gamer*, Issue 003, 2004, 35.

No Man's Sky. Game disc, developed by Hello Games. Guildford, UK: Hello Games, 2018.

Nygren, Niklas (a.k.a. "Nifflas" of Nifflas Games). Interview with author, December 3, 2017.

Nystrom, Robert. *Game Programming Patterns*. Genever Benning, 2014, https://gameprogrammingpatterns.com.

Peggle 2. Mobile, developed by PopCap Games. Redwood City, CA: Electronic Arts, 2013.

Pokemon Blue. Game cartridge, developed by Game Freak. Kyoto: Nintendo, 1996.

Bibliography

Polytron. "FEZ," published 2012, accessed October 10, 2019, http://fezgame.com.

Pong. Arcade, developed by Atari. Sunnyvale, CA: Atari, 1972.

Portal 2. Game disc, developed by Valve. Bellevue, WA: Valve, 2011.

portal2soundvideos, director. "Portal 2: Interactive Music." YouTube, January 4, 2012, www.youtube.com/watch?v=ursIj59J6RU&t=324s.

Puck Man (*Pac-Man* in North America). Arcade, developed by Namco. Tokyo: Namco, 1980.

Raine, Lena (composer of *Guild Wars 2* and *Celeste*). Interview with author, March 1, 2019.

Rapaille, Clotaire. *The Culture Code: An Ingenious Way to Understand Why People Around the World Live and Buy as They Do.* New York: Random House, 2006.

Red Dead Redemption. Game disc, developed by Rockstar San Diego. New York: Rockstar Games, 2010.

Rez. Game disc, developed by United Game Artists. Tokyo: Sega Games, 2001.

RIAA. "Gold & Platinum" Search Portal, "Buckner & Garcia—Pac-Man Fever," accessed December 16, 2019, https://www.riaa.com/gold-platinum/.

Rice, Robert (VR audio specialist). Interview with author, May 2, 2019.

Rice, Robert, and Kedar Shashidhar. "The Role of Audio and Multimodal Integration for New Realities." AES New York 2018 (convention lecture), Jacob K. Javits Convention Center, New York, August 20–22, 2018.

Riley, Terry. *In C.* New York: Associated Music Publishers, 1964.

Rise of the Tomb Raider. Game disc, developed by Crystal Dynamics. Tokyo: Square Enix, 2015.

Rosetti, Gregory. "RPG Town Themes: Evoking Place and Cultural Identity through Music." GameSoundCon (convention lecture), Millennium Biltmore Hotel, Los Angeles, CA, October 9, 2018.

Sellers, Mike. "Systems, Game Systems, and Systemic Games." *Gamasutra* (blog), May 18, 2015, www.gamasutra.com/blogs/MikeSellers/20150518/243708/Systems_Game_Systems_and_Systemic_Games.php.

Sinclair, Jean-Luc. "Immersion, Audio, and Virtual Reality." The XR Date: Música (convention lecture), April 4, 2019, Espacio Fundación Telefónica Madrid, Madrid, https://youtu.be/kGcNQh9SgEo.

Skinner, Jon. *Sublime Text* (3.2.2). Sublime HQ, 2019.

Sound Sky. Mobile, developed by Highkey Games. Cupertino, CA: Highkey Games, 2019.

Space Invaders. Arcade, developed by Taito. Tokyo: Taito, 1978.

Spyro the Dragon. Game disc, developed by Insomniac Games. Tokyo: Sony Computer Entertainment, 1998.

Star Wars: Episode I—The Phantom Menace. Directed by George Lucas. Los Angeles, CA: Lucasfilm/20th Century Fox, digital, 1999.

Summers, Tim. *Understanding Video Game Music.* Cambridge: Cambridge University Press, 2018.

Super Mario Bros. Game cartridge, developed by Nintendo Creative Department. Kyoto: Nintendo, 1985.

Super Mario Odyssey. Game cartridge, developed by Nintendo EPD. Kyoto: Nintendo, 2017.

Super Mario Sunshine. Game disc, developed by Nintendo EAD. Kyoto: Nintendo, 2002.

Super Smash Bros. Melee. Game disc, developed by Hal Laboratory. Kyoto: Nintendo, 2001.

Sweet, Michael. *Writing Interactive Music for Video Games: A Composer's Guide.* Upper Saddle River, NJ: Addison-Wesley Professional, 2014.

Texas Instruments. *Advanced Circuits—Type SN76477 Complex Sound Generator: Bulletin no. DL-S 12612*. Texas Instruments, 1978.

The Last of Us. Game disc, developed by Naughty Dog. Tokyo: Sony Computer Entertainment, 2013.

The Legend of Zelda. Game cartridge, developed by Nintendo EAD. Kyoto: Nintendo, 1986.

The Legend of Zelda: A Link Between Worlds. Game cartridge, developed by Nintendo EAD. Kyoto: Nintendo, 2013.

The Legend of Zelda: Breath of the Wild. Game cartridge, developed by Nintendo EPD. Kyoto: Nintendo, 2017.

The Legend of Zelda: Link's Awakening. Game cartridge, developed by Nintendo EAD. Kyoto: Nintendo, 1993.

The Witcher 3: Wild Hunt. Game disc, developed by CD Projekt Red. Warsaw: CS Projekt S.A., 2015.

The Worst Grim Reaper. Download, developed by Moon Moon Moon. Utrecht: Moon Moon Moon, 2020.

Thomas, Chance. *Composing Music for Games: The Art, Technology and Business of Video Game Scoring*. Boca Raton, CRC Press, 2017.

Thompson, Jerry, director. *Sonic Magic: The Wonder and Science of Sound*. CBC Television, 2016.

Thumper. Game disc, developed by Drool. Providence, RI: Drool, 2016.

Total Annihilation. Game disc, developed by Cavedog Entertainment. New York: GT Interactive Software, 1997.

Trifon, Brian, and Brian White. "Beyond the Presets." GameSoundCon 2018 (convention lecture), Millennium Biltmore Hotel, Los Angeles, CA, October 9, 2018.

Uematsu, Nobuo. "One-Winged Angel." On *FINAL FANTASY VII: Original Sound Track*, SQUARE ENIX, 1997, MP3, accessed December 13, 2019, https://open.spotify.com/track/02hiFgacAzz7zGl5EF53eo?si=u_K90TO3SKOx4hJvzkR_Dw.

Unity. Unity Technologies, 2019.

Unreal Engine 4 (4.24). Epic Games, 2019.

Uurnog Uurnlimited. Download, developed by Nifflas Games. Stockholm: Raw Fury, 2017.

van Elferen, Isabella. "Analysing Game Musical Immersion: The ALI Model." *Ludomusicology: Approaches to Video Game Music*. Equinox, 2016.

Vreeland, Rich. "Music Workshop—FEZ." YouTube video, 44:22, November 20, 2014, https://www.youtube.com/watch?v=PH04VJ8jxvo.

Vreeland, Rich. "Philosophy of Music Design in Games—FEZ." YouTube video, 48:31. April 20, 2013, https://youtu.be/Pl86ND_c5Og.

Weir, Paul. "The Sound of 'No Man's Sky.'" Game Developers Conference 2017 (convention lecture), Moscone Center, San Francisco, CA, March 3, 2017.

Where Shadows Slumber. Mobile, developed by Game Revenant. New York: Game Revenant, 2018.

Whitmore, Guy. Interview with author, April 23, 2019.

Wipeout 2097 (*Wipeout XL* in North America). Game disc, developed by Psygnosis. Liverpool, UK: Psygnosis, 1996.

Wwise (2019.1.6.7110). Audiokinetic, 2019.

Zicarelli, David. "Max MSP." 8.1.2, Cycling '74, 0AD, San Francisco, CA, January 28, 2020.

INDEX

AAA. *See* gaming
Abramson, Zach. *See* industry professionals
aesthetics:
    film music, 28
    game audio, 27–8, 34, 55, 58, 94
    low-bit, 27–36, 39, 79, 85, 98–9
    musical sound effects, 69
    sonic codes as, 64–5, 73
Alcorn, Al. *See* industry professionals
algorithmic music, xviii, 39, 96, 108–9, 137, 139–44, 147–50, 166, 262
    advanced systems, 149, 160
    aesthetics of, 146
    aleatoric, 140–1
    designing, 131, 137, 140–1
    form of, 155
    game data-driven, 138–9, 156
    historical, 139–40
    modularity and, 140
ambience, 4, 8, 12, 46, 75, 115, 139, 186, 223
    designing, 47, 73, 75, 94, 98, 188–9
    in game audio engine, 186, 244
    as part of score, 75
    real-life, 56
    virtual, 171
Anderson, Martin Stig. *See* industry professionals
AR. *See* extended reality (XR)
Asteroids. See video games
audio effects:
    digital signal processing (DSP), 186, 243–4
    equalization (EQ), 84, 139
    flanger, 98
    in FMOD, 216
    reverb, 17, 23, 155, 165, 187, 243
    tremolo, 80
audio implementation:
    audio system, 4, 103, 123, 130, 144, 146, 163, 189, 221
    hardware:
        audio budgets, 34–5, 240, 242, 245, 247
        buffer, 56
        frame rate, 213–14, 223
        graphics, 35, 163–4, 242
        memory, 34, 38, 183, 195–6, 231, 239–44
        in mobile gaming, 163, 242
        optimization, xviii, 231, 239–40, 243–4
        streaming audio, 240, 242, 244, 251
    Musical Instrument Digital Interface (MIDI), 86, 143–4, 154–5, 225
    music systems, 3, 19, 22, 24, 125, 153
        advanced, 23, 112, 122, 149–50
        algorithmic, 138, 144, 150, 153, 156
        designing, 4, 94, 123–4, 133–4
        Dynamic Percussion System (DPS), 142–3, 145–9
        emergence and, 121, 130, 134
        game data and, 19, 123–4, 128–9, 149, 155, 220
        hierarchical, 130
        immersion and, 105
        in middleware, 226
        modular, 150
        Ondskan, 149–50, 156–60
        reactive, 11, 109, 150
    pre-rendered audio, 15–17, 22, 36, 86, 109, 137, 139, 149, 189
    structures, 125
        horizontal, 1, 20–1, 23–4, 124–6, 149, 198, 219, 226
        loops, 11, 16–17, 20–1, 23, 30, 32, 132–3, 138, 140, 186, 190, 215, 218–19
        modular, 126, 129, 149, 155
        stingers, 1, 19, 21, 55, 62–3, 125, 217, 225–8
        transitions, 17–21, 24, 125, 145, 214, 221, 225, 227
        vertical, 22–3, 113, 130
    in VR, 162–9
    See also extended reality (XR)
augmented reality. *See* extended reality (XR)

Barnes, Craig. *See* industry professionals
Bendy and the Ink Machine. See video games
Benis, Mark. *See* industry professionals
BioShock. See video games
Brown, Daniel. *See* industry professionals

Call of Duty. See video games
Celeste. See video games
Chang, Larry Yucheng. *See* industry professionals
codes:
    aesthetic conventions within, 27–8, 31, 60–2, 65, 94, 257
    creating new, 53, 76
    cultural, 51, 53–4
    film music, 28, 52–3
    marketing with, 51–2, 54–5
    sonic and musical, xviii, 6, 9, 30, 48, 51–65, 67, 69–70, 76–8, 171–2
    universal, 51, 53, 55
    using pre-existing, 30, 58, 61, 63
    of video game culture, 27, 29, 37, 54–5, 57, 62
Cox, Stephen. *See* industry professionals

Dead Space. See video games
Desiderio, Erik. *See* industry professionals
diegesis, 5, 105, 108, 112–13, 130
    diegesis-nondiegesis continuum, 5, 92, 105
    diegetic music, 65, 92, 110, 170
    enhancing immersion with, 113–14, 172
    in film, 64
    player-music interaction and, 103, 105, 108, 110–13, 130

Index

diegesis (*cont.*)
    sonic relationships and, 92–3, 99
    sound effects and, 32, 35, 46, 75, 109, 112–13
    in Unity, 186
    in VR, 168, 170–1, 173 (*see also* extended reality [XR])
    world-building with, 58, 73
Disasterpeace. *See* industry professionals
Dolby, Thomas. *See* industry professionals
Donkey Kong. *See* video games
DPS. *See* audio implementation
Duck Hunt. *See* video games
Dynamic Percussion System. *See* audio implementation

Electroplankton. *See* video games
Elias. *See* middleware
emergence, 121–7, 129–34, 149
extended reality (XR), 162
    augmented reality (AR), 162
    mixed reality (MR), 162, 174
    virtual reality (VR), xviii, 161–74, 262, 264
        audio spatialization, 163, 165–8, 170–1, 173
        designing scores for, 168, 170, 173
        gaming market, 161
        hardware, 161–3, 165, 169
        immersion, 162, 168
technical considerations, 161, 165, 168–73

FEZ. *See* video games
Final Fantasy. *See* video games
FMOD. *See* middleware

gaming:
    AAA, xix, 96, 193
    indie, xviii, 96, 149, 160, 177, 193, 247–8, 251
        composer, xviii, 130, 240
    mobile, xviii, 161, 163–4, 174, 189–92, 242, 246, 256, 260, 263–5
generative music. *See* algorithmic music
Giants: Citizen Kabuto. *See* video games
Grand Theft Auto. *See* video games
Graves, Jason. *See* industry professionals
Grobo. *See* video games
Guitar Hero. *See* video games

Hale, Andrew. *See* industry professionals
Halo. *See* video games
Horizon Zero Dawn. *See* video games

indie. *See* gaming
industry professionals:
    Abramson, Zach, 125, 127–8
    Alcorn, Al, 31–2
    Andersen, Martin Stig, 222–3
    Barnes, Craig, 130–3, 135, 261
    Benis, Mark, 221–2, 261
    Brown, Daniel, 142–3, 145, 148
    Chang, Larry Yucheng, 172
    Cox, Stephen, 171
    Desiderio, Erik, 170, 174, 261
    Disasterpeace (Richard Vreeland), 7–8, 14, 29, 55, 94, 96–9, 101, 250, 262, 265
    Dolby, Thomas, 137, 147
    Graves, Jason, 55, 62, 172
    Hale, Andrew, 57
    Kondo, Koji, 28–9, 34, 54, 75, 77, 89
    Lamperski, Phil, 145–6, 148, 263
    Lundgren, Robert, 225–8, 230, 263
    Masuda, Junichi, 11, 14, 263
    Mitchell, Briar Lee, 121
    Mraz, Ronny, 125–7, 129–30, 134–5, 170, 261
    Nifflas (Niklas Nygren), 78, 89, 112, 149, 156–60, 263, 265
    O'Donnell, Martin, 37
    PHÖZ (Noah Kellman and Alba Torremocha), xviii–xix, 17, 73–4, 95, 150–1, 170, 181, 188–9, 249
    Raine, Lena, 29, 133, 135
    Rice, Robert, 161
    Rossetti, Gregory, 8
    Salvatori, Michael, 37
    Sinclair, Jean-Luc, 172
    Taylor, Chris, 133
    Thomas, Chance, 4
    Uematsu, Nobuo, 14, 28, 34–5, 37
    Weir, Paul, 123, 135, 138, 147, 265
Inside. *See* video games

Kellman, Noah. *See* industry professionals
Kondo, Koji. *See* industry professionals
Kontakt. *See* sampling

Lamperski, Phil. *See* industry professionals
Legend of Zelda, The. *See* video games
Lundgren, Robert. *See* industry professionals

machine learning, 141, 147
    Markov chains, 158
    musical phrases, 141
Mario Kart. *See* video games
Mario Party. *See* video games
Mass Effect: Andromeda. *See* video games
Masuda, Junichi. *See* industry professionals
Michael Jackson's Moonwalker. *See* video games
middleware, 17–18, 25, 129–30, 142, 166, 191, 213–15, 217–19, 221, 223–5, 227–9
    Elias, 25, 129, 213–15, 224–30, 262
    FMOD, 25, 129–30, 142, 166, 213–18, 229–30
    Wwise, 25, 123–4, 129, 142, 166, 213–15, 218–22, 224–5, 229–30, 244, 265
MIDI. *See* audio implementation
Mini Metro. *See* video games
Mitchell, Briar Lee. *See* industry professionals
mixed reality. *See* extended reality (XR)
mobile. *See* gaming
MR. *See* extended reality (XR)
Mraz, Ronny. *See* industry professionals
Musical Instrument Digital Interface. *See* audio implementation
musical sound effects, 52–3, 69–71, 76, 78, 94
    action reinforcement, 71, 91, 94
    aesthetics of, 35, 54, 77, 79
    in audio systems, 222
    codes and, 55–6

designing, xviii, 32, 70, 72, 75–7, 79, 81, 83, 85, 87, 139
diegetic, 114
history of, 32, 34–5, 39
item-get fanfare, 39, 71
music-to-sound-design relationships, 48, 95
providing clues, 72, 74, 95
stinger, 19
in VR, 173
music composition:
    algorithmic, 143, 150, 160
    arpeggiation, 33, 39, 58, 77, 109, 153
    arrangement and orchestration, 17, 27–8, 30–1, 34, 36, 39, 58–9, 64, 125, 129, 153
    atonality, 16, 22, 79
    chromaticism, 76–7, 132, 141
    diatonicism, 6, 76
    dissonance, 58, 62–4, 132, 157
    dynamics, 127, 140, 152, 155
    harmony, 6–7, 33–4, 52–4, 60–1, 64, 76–9, 92, 96, 99–100, 132, 153, 157
    leitmotifs, 10, 59, 63, 76
    melodic transformations, 151–2, 154
    melody, 34, 52, 59, 63–4, 73–4, 76, 83, 132, 141, 150–2
    modularity, 140
    nonlinear form, xvii, 1, 3–4, 6, 9–10, 15, 31, 36–7, 146, 155
    phrase, 18, 54, 63–4, 151–2, 157, 218
    rhythm and groove, 23, 52, 54, 56, 60–1, 70, 75–6, 78, 100, 108, 150–1, 154
    sonic codes and, 52
    tempo, 7, 18, 132–3, 154, 218
    timbre, 6, 20, 55, 59, 62–3, 137, 154

Nifflas. *See* industry professionals
Nygren, Niklas. *See* industry professionals

O'Donnell, Martin. *See* industry professionals
Ondskan. *See* audio implementation

Pac-Man. *See* video games
PHÖZ. *See* industry professionals
player experiences:
    actions, 3, 9, 69–70, 91, 96, 99–100, 103–5, 108–9, 111–13, 150, 156
    arc of narrative, 5, 11–12, 146
    emotions, 4–6, 11–12, 45, 61–2, 70, 106, 115, 145
    forming associations, 5–6, 9–10, 12, 29–30, 46, 48, 51–2, 59, 63, 65, 76, 78, 150
    immersion, 38, 45–9, 91–3, 99, 103–5, 109, 112, 114–15, 161–2, 168–73, 222
    interaction with music, xviii, 4, 36, 39–40, 45–6, 48, 52, 100, 103–6, 108–15, 121
    outside of gameplay, 29
    in VR, 172
Pokémon. *See* video games
Pong. *See* video games
proceduralism, xviii, 33, 39, 124, 137–9, 141–3, 145–9
    creating game environments with, 138
    procedural systems, 138, 143–4, 147
    randomization, 132, 140–1, 156, 189
    in Unity, 188–9, 205, 207–8, 213
    See also algorithmic music
programming:
    object-oriented programming (OOP), 195
    programming languages, 142–4, 193–5, 210, 224, 228
pseudocode, 180, 182, 198–9, 210

Raine, Lena. *See* industry professionals
Rez. *See* video games
Rice, Robert. *See* industry professionals
Rise of the Tomb Raider. *See* video games
Rossetti, Gregory. *See* industry professionals

Salvatori, Michael. *See* industry professionals
sampling, 35, 79, 86
    in music systems, 34, 124, 144
    sampled instruments, 34, 62, 86–7, 130, 144–6
    sample rate, 98
    samplers, 29–30, 86–7, 154–5
        Kontakt, 86–9, 263
        round robins, 86
Sensory, Challenge-Based, and Imaginative (SCI) Model, 45–6, 48–9
Sinclair, Jean-Luc. *See* industry professionals
Space Invaders. *See* video games
Super Mario Bros.. *See* video games
Super Smash Bros.. *See* video games
synthesis, 70, 79–80, 85–6
    in early video game consoles, 28, 34, 69, 79, 138
    synthesizer components:
        ADSR envelope, 80–2
        filter, 23, 80, 84–5, 98, 155
        low-frequency oscillator (LFO), 80, 98
        noise generator, 84, 98
        oscillator, 33, 79–80, 82, 85, 98
    varieties:
        additive, 85
        frequency modulation (FM), 85
        phase distortion (PD), 79
        subtractive, 79–80, 82, 85
        vector, 85
        wavetable, 85
    waveforms, 80, 82–3, 85–6, 98, 133, 141, 154, 157
        the overtone series, 82–3
        pulse-width modulation (PWM), 34, 83
        saw, 83
        sine, 55, 83, 85
        square, 34, 83
        triangle, 34, 83

Taylor, Chris. *See* industry professionals
Thomas, Chance. *See* industry professionals
Tomb Raider. *See* video games
Torremocha, Alba. *See* industry professionals

Uematsu, Nobuo. *See* industry professionals
Unity:
    assets:
        clip, 183, 185–6, 196–8, 202–3, 206–8, 224, 240, 244
        container, 140, 183, 185, 218–20

Index

Unity (cont.)
    management of, 239–40, 242, 244
  audio engine, 181–2, 186, 188
  debug, 179–80
  functions:
    Awake, 186, 205–6, 209–10
    Execution Order of Events, the, 202
    PlaySound, 143, 196–8, 206–7, 209–10
    Random.Range, 207–8
  game call, 215, 222, 225, 227
  game objects, 178, 196–7, 201–2, 204–5, 207–10, 222
    AudioClip, 183–6, 196–8, 200–2, 207–8
    AudioManager, 183–4, 205–6, 209
    AudioSource, 170, 183–7, 191, 197, 204, 206–8, 244–6
    gameObject, 205–6, 208–9, 224
    MusicManager, 189–91, 244
    MusicSource, 184–5
    public, 183
    Rigidbody, 201–2
    singleton, 205, 209–10
  user interface (UI), 178–9, 183, 185
    Console, 180
    Inspector, 179, 184–5, 200
    Profiler, 243
Unreal Engine, 166, 177–8, 227–8
Uurnog Uurnlimited. See video games

version control, 231–8
video game mode, 37–8, 226–7
video game music business:
  budgets, 32, 229, 247–8, 252–3, 255–6
  compensation, 247–53, 258–9
  composer responsibilities, 258
  contracts, 247–53, 257
  deadlines, 248–50, 253, 258
  "exposure", 260
  intellectual property (IP) rights, 249, 251, 258–9
  negotiation, 249–50, 253
  networking, 247, 254–7
  rewrites, 258
video game music functions:
  character themes, 9–10, 15, 155, 171
  loading screen music, 38
  locational themes, 6–9, 12, 37, 60, 83–4
  mini-game music, 38, 133
  overlap with musical sound effects, 7, 38–9, 71
  pause screen music, 38, 191
  player-music interactivity, 31, 36, 46, 74, 172
  situational themes, 11, 37, 39, 155, 226–7
  title screen music, 32, 37–8, 252–3
  world-building, 73–4, 76, 91, 95, 188
video games:
  Asteroids, 27, 32, 40, 261
  Bendy and the Ink Machine: Chapter Two: The Old Song, 111, 116, 261
  BioShock (series), 65, 67, 261
  Call of Duty (series), 46, 49, 93, 261
  Call of Duty: WWII, 104
  Celeste, 29, 41, 135, 214, 230, 261, 264
  Dead Space (series), 55, 62, 79
  Donkey Kong (series)
    Donkey Kong, 54
    Donkey Kong Country, 75
  Duck Hunt, 77
  Electroplankton, 109, 262
  FEZ, 91
  Final Fantasy (series), 10–11, 14, 37, 42, 262, 265
    Final Fantasy, 37
    Final Fantasy VI, 35
    Final Fantasy VII, 10, 37
  Giants: Citizen Kabuto, 35, 42, 262
  Grand Theft Auto (series), 31
  Grobo, 17, 150–2
  Guitar Hero (series), 108, 131
  Halo (series), 11
    Halo 5: Guardians, 37
    Halo: Combat Evolved, 37
  Horizon Zero Dawn, 58, 105–7, 112, 122, 135, 262
  Inside, 113–14
  Legend of Zelda, The (series), 39, 42, 54, 71, 89, 101, 263, 265
    Breath of the Wild, 78, 91, 110–12
    Legend of Zelda, The, 37, 83, 97
    Link Between Worlds, A, 75
    Link's Awakening, 71–2
  Mario Kart 64, 6, 14, 60, 70, 89, 263
  Mario Party (series), 38–9, 42, 263
  Mass Effect: Andromeda, 221, 230, 263
  Michael Jackson's *Moonwalker*, 30–1
  Mini Metro, 55–6, 92, 96, 99–101, 261–3
  Pac-Man, 34, 36–7
  Pokémon (series):
    Pokémon Go, 174
    Pokémon Red and Blue, 6, 11
  Pong, 31–2, 41, 69, 177, 192, 264
  Rez, 108, 264
  Rise of the Tomb Raider, 145–6
  Space Invaders, 32
  Super Mario Bros. (series):
    Super Mario 64, 10, 14, 261
    Super Mario Bros,, 28, 32, 36, 54, 76–7, 83–4, 91
    Super Mario Odyssey, 29
    Super Mario Sunshine, 29
  Super Smash Bros. (series), 38–9, 42, 263
  Uurnog Uurnlimited, 78, 89, 112, 149, 156, 160, 265
  Worst Grim Reaper, The, 221, 230, 261, 265
virtual reality. *See* extended reality (XR)
VR. *See* extended reality (XR)
Vreeland, Richard. *See* industry professionals

Weir, Paul. *See* industry professionals
Worst Grim Reaper, The. *See* video games
Wwise. *See* middleware

XR. *See* extended reality (XR)